50 DEGREEING A CAM
Get the most out of the cam by degreeing it.

54 PISTONS - RINGS - RODS
Smart assembly steps.

58 COMPLETING THE BLOCK
Oil pan science, picking the right oil pumps, and leak-proofing.

62 HEAD PREP
Everything you need to know from valve jobs to final assembly. Also being able to run on unleaded.

68 TOP END ASSEMBLY
Head and valve train.

72 INTAKE TECH
There is bolt-on horsepower and torque in your manifold choice.

75 INTAKE BOLT ON
Make room for larger ports and prevent leaks.

78 THE GREAT COVER UP
Your choice of valve covers, plus good sealing.

82 FINAL ASSEMBLY
Putting all the finishing touches on your pro engine.
| STARTER | FUEL PUMP | CARBURETOR | START UP |
| EXHAUST | DISTRIBUTOR | Install and Tune | THE LITTLE BIG THINGS |

88 BIG INCHES
It's colossal - Godzilla has landed - can you handle 639 cubic inches?

94 457 CUBIC INCH SMALL BLOCK
Switch to GM's Rocket small block and find the newest look in cast iron.

97 FACT FINDER
Clearances and torque values

98 ADDRESSES
A handy list of people who help you go fast.

Our new cover is that of a Winston Cup Chevy engine, built by Bob Rinaldi at Moroso's Race Engine Shop. Director of Research, Fred Golja helped us set up the shots during final assembly. Fred and Bob are responsible for much innovative dyno testing and advanced components.

LISTEN TO THE HEARTBEAT OF CHEVROLET

HERB FISHEL, Motorsports Technology Group

A Chevrolet will win races on gasoline, alcohol, with carburetors or fuel injection. A Chevrolet is a winner's car. Both Small and Big Blocks dominate the race tracks. "A guy who consistently wins makes his own luck."

"Don't trick yourself out of the race: keep it simple and kick ass! We are going to assemble a bullet proof engine, to make sure the Chevy up front is yours."

Alex Walordy

ALEX WALORDY PRESS
P.O. BOX 623
Westbury, NY 11590
PHONE 516 ● 334-8365 - FAX ● 516 997-0881

ISBN #0-9624323-0-X

LIBRARY OF CONGRESS
CATALOG #89-090429

ENGINE BUILDER'S CORNER

There is beauty in an engine that you have built yourself. It sparkles like no other when you install it in your engine compartment. Attached to it are things you cannot bolt on, like a feeling of pride and achievement. On completion there will be more money left in your wallet, a brighter gleam in your eye, and two long black tire streaks at the track.

John Lingenfelter, Earl Gaerte, Ron Shaver, Reher-Morrison, R.H.S., the hobby stock racer around the corner, the young fellow down the block who is assembling his first engine—all share something in common: they are engine builders. From Ground Zero to Level 10, they share a natural curiosity, a desire to tackle the new and the unknown plus a will to win.

The big name guys share another thing with the little guys—for the most part they started out small and worked up. We did our first story with Jim Cavallaro of Diamond Racing in a narrow little Detroit shop and our first one with John Lingenfelter at a little motel in Pomona during a Winternationals, too long ago. At the start, C.J. Batten had just quit an engineering job and was grinding heads in his garage. The partners who began R.H.S. were local Memphis, TN racers. Today, all of them know every nut, bolt and computer on the machines in their huge shops and the one thing in which they are superior is gathering knowledge. Come join their elite engine building club.

Our *Engine Assembly Guide* is designed to help you find out about engines quickly so you can get off to a good start. It will show you what to look for, give you the expert's understanding of design points and make you aware of a thousand details as you take the engine apart or put it together. It shows you how to precheck and prefit so once the engine is ready to go together, you will not be involved in last minute do overs. You will save many times the initial cost of the *Engine Assembly Guide*, just from having a feel for the right combination and the good engine parts. It is just like having a panel of experts at your side going over the main points step by step so your engine will make more power and be more driveable.

Instead of throwing a fistful of money at the project, government style, you get more horsepower per dollar. You are now the inspector who knows how to check out the options and the pieces.

CAUTION-WARNING-CAUTION

We want you to build the best possible engine, without spending a mint for it. This book has been assembled to help you in this endeavor. To do it, we have drawn on the brain power of some of the best engine builders in the country from East to West and from South to North. It has been read, checked, and rechecked. However, Murphy's Law holds and **THERE MUST BE ERRORS IN THIS BOOK.**

When you build an engine there is only one guy responsible for it, and that is you. The man doing the final assembly is the final inspector. We could tell you the moon is made of blue cheese—you are the final authority, and the only judge. If you see a real boner, call us and let us know. It will help the next edition.

If you build the engine, then lean it down, bump up the spark, raise the compression, and decide to recoup your costs with a load of cheap gas we'll be glad to extend our sympathy, but just remember we told you not to do it.

Every racer in the U.S.A. has spent a great deal of time studying broken pieces, trying to figure the cures. That's what made him a better racer next time around. Some people are easy on equipment, others can break an anvil—some learn, others never do.

Here is the best possible info, prepared with care. Use it wisely and the end result is up to you. **THIS IS A LEGAL DISCLAIMER FOR ALL PROBLEMS, REAL AND IMAGINARY, INCURRED FROM READING THIS BOOK.** However, it is also there to turn on a light bulb and say the only way to learn to build engines is to be careful, read all the info you can lay your hands on, and also turn on the bullshit filter, decide what you think is right or wrong, and hold your own counsel.

Alex Walordy

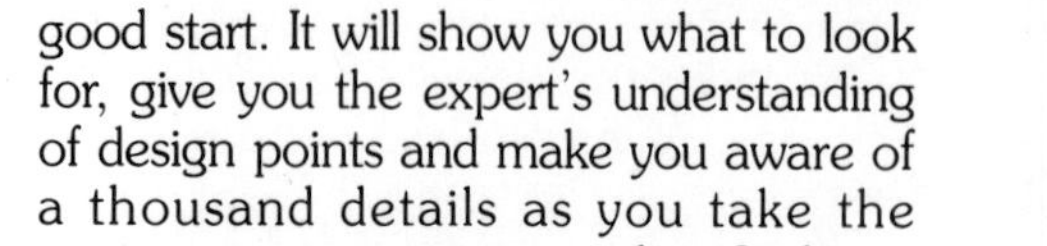

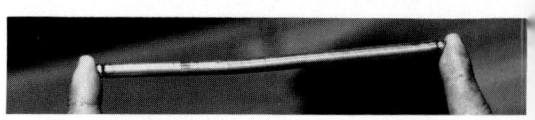

You could shrug shoulders and say: it happens—or you can work at preventing it.

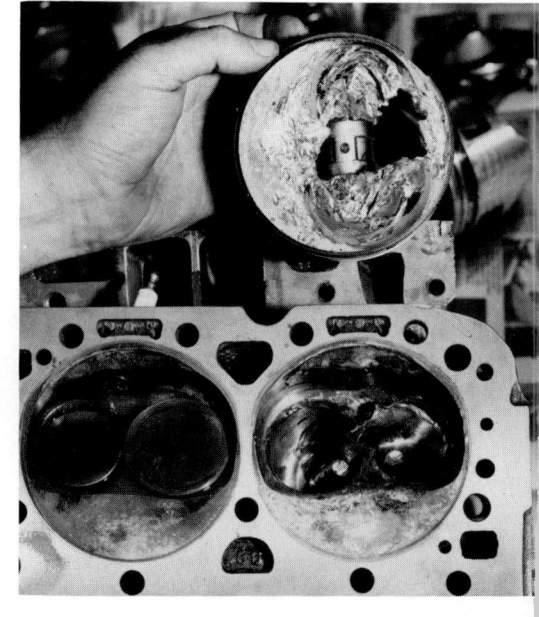

MORE INFO

HOW TO USE THIS BOOK

If you have the time, the best way to use this *Engine Assembly Guide* is to read it cover to cover. Lacking that, just look at the pictures and read the captions. We made it with the best, clearest pictures possible and they are each worth a thousand words. If you are short of time, just refer to the one section that pertains to you, be it installing a cam, bolting on a head, or fitting in a starter.

If you plan to do cylinder head work, you'll find tons of info in the HEAD PREP, COMPRESSION AND GASKET sections.

TORQUE AND POWER COMBINATIONS tells you how to gain more of each. Now your engine planning becomes easy. You are the conductor and all the players in your combo are playing the same tune. We have made each chapter complete and self sufficient, which has caused a minor overlap here and there, but it speeds a quick look up.

BEST INFO

You would hardly drive across country without a map and building an engine without a shop manual is almost as bad as stopping at each street corner for directions. There is little to equal the factory shop manual in terms of pages, photos, and information. It is not going to show you how to build a high performance engine or a race motor, but it will carry all the factory specs, the tools for the job, and other indispensable information.

All Chevy manuals are available from Helm, Inc.

Publications—Customer Service
14310 Hamilton Ave.
Highland Park, MI 48203
313 865-5000

Also available from them are Electrical and Vacuum Trouble Shooting Manuals.

HOLLEY CARB INFO

SUPER TUNING HOLLEY CARBURETORS

This practical book tells you how to tune the carburetor to win races. It goes through all the metering block drillings, track testing, machining, and bench work. You will find out how to gain more air flow, higher torque, and more top end power.

ALEX WALORDY PRESS
PO Box 623
Westbury, NY 11590
FAX 516-997-0881
Telephone 516-334-8365

Another excellent source of info is the *Chevy Power Book* and the *Performance Parts Catalog* available at any dealership.

You get a monthly update by reading good tech oriented magazines such as *Popular Hot Rodding*, *Super Chevy*, and *Off Road* put out by Argus Publishers and Petersen's *Circle Track*, *Hot Rod*, and *Car Craft*. Lopez Publishing puts out *Stock Car Racing* and *Super Stock*.

What you get from this ENGINE ASSEMBLY GUIDE is invaluable information on components and how to sort out what you see. You also gain specific performance engine assembly info, as though a team of engine builders was there in your engine shop to help you build it.

Rick Brown from Lingenfelter Racing Engines makes a science out of selecting the best combinations from box engine kits—a way to get the most power per dollar.

help is Chevrolet Raceshop, people like Herb Fishel, Dick Amacher, Roger Allen, Jim Covey, the list is long. Bill Howell, now retired and busier than ever, was instrumental in keeping us up to date on the best of Bowtie parts. CPC (Chevrolet—Pontiac—Canada) comes up with the right pieces and are a main spring.

A book is dead in the water without pictures and four shops who really threw their doors open with no accounting for hours are: Dick Titsworth of Seaport Automotive, Lee Bandrow of LAB Machine, Bob Patzold of Prototype Engineering, and Keith Burton at Jeg's. Lee teamed up with Lou D'Amico to put

Earl Gaerte is putting the finishing touches on a new engine. After a dyno break in, the heads and all bolts are retorqued. Now the fine tuning begins, with the dyno providing the answers. Injectors spray at the intake and into the valve bowl.

THANKS

People have asked us with which engine builder we did this book. It would be easy to just put down a name. Truth of the matter is that we have received engine builder help and information from hundreds of sources and listing them all, including the many racers who build their own, would outstrip these pages. Contrary to trick of the week secrets, engine builders and racers are amazingly open and helpful. Yes, if you were running against them on Saturday night at their home track, they might be a little more cagey but, by and large, they know that they can build the best engine, love engines and don't mind discussing them. Manufacturers of parts, tooling, and cars are a live source of new information.

Probably topping the list for editorial

INFO • 5

It takes magic fingers on a computer to make a book come about. If you want to enter it in really fast how about two keyboards? Meet Nancy—she did it all!!!!

together a very fully instrumented dyno and every year they come back from the Detroit session of SAE with just a few more gadgets...

Then too, Scooter Brothers of R.H.S. and Tom Woitesek of Competition Cams provided countless man hours of hands on shop and engineering discussion. Cal DeBruin of Speed-Pro combines some sharp engineering with a practical racer outlook. Gary Wade, now at Crane Cams, is always there to answer questions or to meet us on a short notice at Earl Gaerte or John Lingenfelter's. Bill Mitchell who has more NHRA records to his name than most, has just finished designing a new set of small block cast iron heads followed by big block heads and aluminum small blocks. In his spare time he builds big inch drag race motors and runs a short track open wheeler at Riverhead, NY.

Getting out a book without a computer is like horse and buggy days. We wanted some little extras, like two screens and two keyboards, and, and, and,... All the experts told us why it was impossible and then Greg Rieber, the Katech electronics man, pointed us in the direction of Alan Darge of Sun Data who said "no trouble" and solved everything the easy way. Now we want you to know a computer does not work by itself and yours truly was not about to volunteer so Nancy Walordy learned the contents of umpteen thick books and put in every single story you see in here—also laid out most of the pictures.

The pictures are all taken with an RB 67 Mamiya camera, and the large assortment of lenses that go with it. It's a large format piece, bigger and more powerful than a Hasselblad. The RB has five square inches of negative area, as opposed to one square inch on a 35 mm, and on it are mounted two moose antler headlights that have blinked at everybody in the industry. Then too, there is Greg Rogers who gets the credit for processing every negative and print used in this book.

"Hands-on" is the most important part of this book, from the early days of Don Garlits' Swamp Rat IV to Gaerte's latest Late Model to building our own street engines, to spending unlimited hours at the Batten Heads tooling center filled facilities.

We always felt that Big Daddy Garlits' biggest secret was that he worked on his race car from the time he got to the track, to the time he left and then worked on it again from the time they hit the motel, to some time the next morning.

Look for little details that make the big difference. Here, we receive priceless help from Keith Burton at Jeg's in Columbus, OH as we watch every step of his putting an engine together and at Seaport Automotive where Dave Jascob watched every step of our putting an engine together—and both he and Dick Titworth came up with invaluable tips and hints.

Down in Stone Mountain country in Decatur, GA, Ralph Thorne found time to go through all our photos and contributed shop time for us to gather more of them. Ralph has a knack for boiling down complex thinking into a real world answer.

Just around the corner from there, cam grinder and carburetor builder John Reed has a great grasp of the total engine, in addition to his incredible computer talent. Our piston section involved the combined assistance of John Whittaker from Zollner Pistons, Jim Cavallaro of Diamond Racing Engines, and Bill Willett of Speed-Pro.

Jerry Rosenquist and Roger Friedman of Fel-Pro combined hands-on experience with considerable design talents and field knowledge to help us with the gasket section.

An amazing amount of refinements go into an intake manifold and even more testing is involved in matching carburetors to the engine. The engineers at Holley from Bob Miller to Jim Guibord and Ralph Bugamelli helped us make it an open book.

Then we visited Gene Ohle at Evans Machine in El Monte, CA and spent more time learning block machining tricks. Now you know why this book took a year instead of a month.

Jody Schmeisser and Dave DeVito of S/D Race Cars do their own engine assembly. Prototype gets credit for all the machining. Both Fel-Pro and Speed-Pro tech services have extended much assistance and have used this fine machine as a test bed for performance products. All main engine components are Bowtie.

You pull on the wrench and apply torque to a bolt. By the same token the wrench handle can turn and apply a FORCE. Let's trace through what happens to engine torque as the ring gear turns the rear wheels. The torque is operating through an arm—the radius of the wheel and tire. The force at the end of this arm pushes back against the ground. Think of a wheel as an infinite series of those torque arms, ready to transfer force to the ground.

The ring and pinion multiply the torque to the rear wheels. Your car accelerates faster with a higher number gear, just as though you had put a bigger engine in the car. Also peaks out earlier.

6 • INFO

CHEVY POWER PARTS

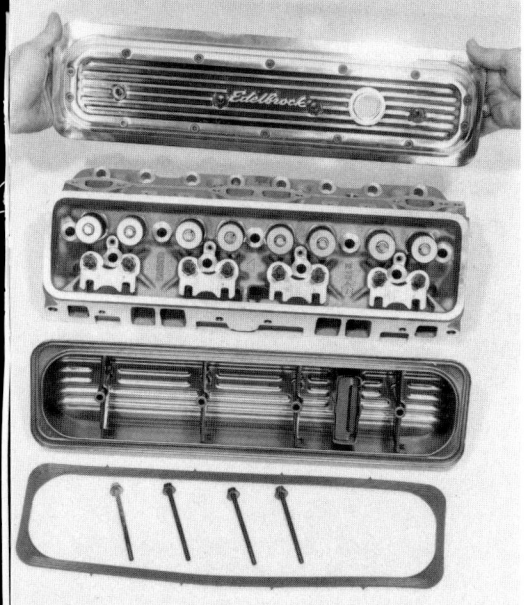

Conventional valve covers are retained by screws engaging the cylinder head rail, which requires load spreaders. On an aluminum Corvette head, rocker covers with four central studs achieve superior sealing.

At one time they kept deep secrets from each other. Chevy wasn't supposed to know what Pontiac was doing and Buick fought as hard with Olds as Chrysler does with Ford. Today all of the top racers have been brought in from the divisions to work as a team with Herb Fishel and Motorsports Technology at the GM Tech Center. They have a common goal: win races for GM instead of fighting division against division. It's like no longer

Chevrolet's new canted valve head has the intake and exhaust ports separated and widened for max flow. You will need a pro style sheet metal intake manifold and valve covers.

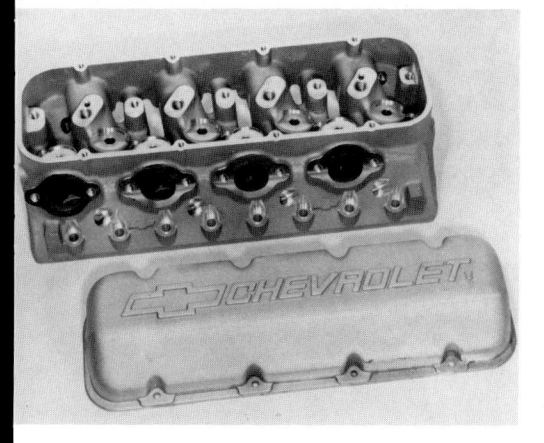

scoring goals against your own teams. Now, you can now buy these powerful Motorsports parts from the Chevy dealers, which is exactly what this story is all about.

We asked Nick Fowler of Scoggin-Dickey Chevrolet in Lubbock, Texas to bring us up to date on the very latest in Chevy Power. They supply racers all over the United States and know early the "best" parts in the class you run. Nick is a racer, an engine builder, and has an uncanny knowledge of pieces and applications, a wide world that ranges from street to dragsters and oval track. Here are some of the latest and most useful Chevy Power Performance Parts.

CAST IRON STREET

The cast iron 10105123 block is perfect for a moderate rpm street engine - also very affordable. Lifter bores on this block are 0.630 inch taller than on the old block to better accept hydraulic, mechanical billet lifters or GM hydraulic roller lifters. The extra height allows GM to use their hydraulic roller lifter retaining bars. Taller bores also provides better guidance and oil pressure retention for the hydraulics.

Converting this block to mechanical roller lifters would require extensive machining, because the extra lifter boss height interferes with mechanical roller lifter line-up bars. Aftermarket racing roller lifters are designed for the older, lower block and no one currently makes or plans racing roller lifters for this block. Use the 10105123 for what it is, a hydraulic street block.

Scoggin-Dickey makes a rear seal adapter for the late model cast iron street block, so you can use the stronger early style crank, with its larger flywheel flange. Another benefit of the adapter is that you can save money by retaining the original oil pan and windage tray. The new pan with its one piece seal is great and its molded one piece gasket eliminates oil leaks. However, the gasket and pan are expensive, so the rear seal adapter conversion makes money sense.

BOW TIE BLOCKS

To gain more horsepower you need rpm and now you also want a stouter block. Bow Tie blocks are all very much upscale from the cast iron stock replacement block. They differ in the main bearing caps. The first is the Bow Tie block 24502458 which carries production four bolt gray iron caps. They are sized for 350 mains and will accept all the pieces for the 350 or 400 block, in other words, a good street motor for modest money. There are also Bow Tie blocks with two bolt mains so you can use the caps of your choice and machine the block as you want.

Next up is the 10185047 which carries four bolt nodular iron caps and is good for any compression application under the 600 hp level. The mains are bored to a 350 size, but all bolts head straight in.

Splayed cap Bow Ties are a must above the 600 horsepower mark and preferred for longer racing. They have the two center bolts heading straight up and the two outer bolts angled out,

Giant intake ports and canted valves allow you to make full use of a fabricated sheet metal manifold. All four ports are "good".

CHEVY POWER PARTS • 7

Molded one piece oil pan gasket, designed for the late blocks and one piece seals, leak proofs the engine.

Unlike the 492 in the center, the Bow Tie does not use an exhaust crossover. Extra metal strengthens the intake port area. Pushrod holes can move further apart.

which provides better block tie-in. There are Bow Tie part numbers for wet sump use where the rear cap accepts a stock pump, or for dry sump motor with a solid cap and differently drilled oil galleries.

350 VS 400 JOURNALS

A 400 crank with its larger journals offers strength benefits, particularly with a stroker crank. It is beneficial in a tight revving, high horsepower engine. The 350 journal sizes are smaller and so the linear velocity (rubbing speed) of the bearing and crank is lower and friction horsepower is reduced. The 350 rod journals are smaller which allows more room for strengthening the bottom end of the rod. For durability in a 7000 rpm and up stroker, go to 400 journal sizes.

IT'S ALL IN THE HEAD

A basic 492 cast iron straight plug head (3987376) meets sanctioning body requirements in lower classes and is also fine for strong street torque. Among the specs, you'll find 2.02 intakes, 1.600 exhausts and a moderate 157 cc runner volume. This gains ample port velocity for the low and mid range and gives traffic throttle response. A 492 head is shipped from GM with guide plates and screw in studs. The chambers come in at 64 cc so you can build compression.

Bow Tie heads all come with angle plugs and the new Phase 2's have improved combustion chambers which also come in at 64 cc, just like the 492. Earlier Bow Ties had 67 cc heads. Among the Bow Tie Phase 2 plus points are a thicker deck and considerably improved exhaust ports. Also, the intake ports carry a 190 cc runner volume. The experts at Scoggin-Dickey feel

The Bow Tie block carries the heavy duty Bow Tie emblem. It comes through with unfinished bores, machined to the size you choose.

The lower half of the Scoggin-Dickey two piece rear seal is retained by four Allen bolts. Installed on a late model block, it allows you to retain the old style rear pan and crank.

8 • CHEVY POWER PARTS

Combustion chamber of the angle plug Bow Tie at the right is tighter, saves cc and increases squish area to help the engine stay out of detonation.

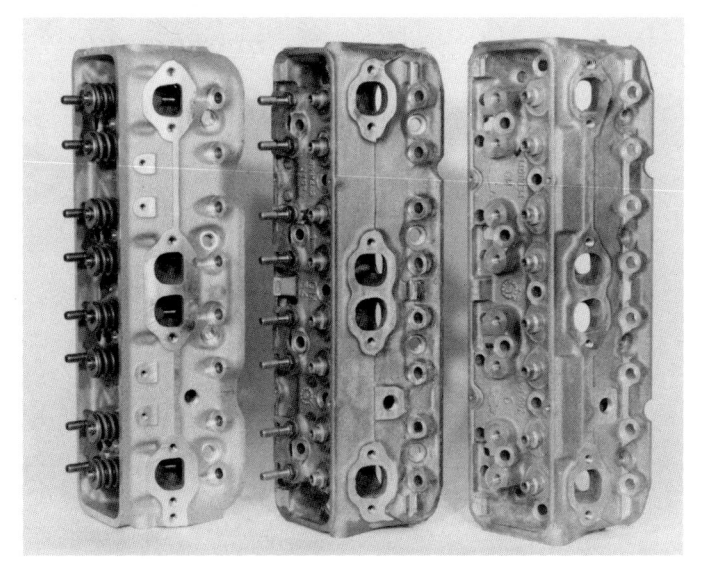

At the left, the aluminum 350 HO Corvette head, in the center a straight plug 492 head, required by many of the racing associations. High powered, cast iron Bow Tie, angle-plug head is at the right.

that this is good volume for any 350-400 cubic inch engine up to a 6800 rpm range. Bow Tie heads are shipped bare and accept guide plates and screw in studs. You can also get them performance assembled.

18 DEGREE HEADS

The 18 degree head does make more power. Porters have started to find all the sweet spots and the horsepower curves point up. It represents the latest in GM technology but is also priced at double the Bow Tie tag. In Winston Cup they are "the" head and tie into the Bow Tie intakes.

CORVETTE ALUMINUM HEADS

The understandable street trend is to use the aluminum Corvette heads. They are light and the aluminum also conducts heat well so they are less sensitive to octane ratings. It has a flat style chamber, angled plugs, and the intake port volume is 163 cc. Fact is that this head produces outstanding torque numbers and incorporates the latest in swirl technology. If anything, it requires less spark than a conventional head. Scoggin-Dickey racers see really good results at 32-34 degrees total advance. You get good response to camshaft changes in the 214 to 224 degree range.

The valves are smaller than in a Bow Tie, 1.94 intakes and 1.50 exhausts. Since it is an aluminum head, it comes with factory installed seats which cannot be reworked for big valves. Replacing them can be done, at extra expense. In effect, it was not built for big valves. Conversely, for big valves there are many Chevy aluminum heads that are very competitive. Swirl action

Dowels align the new rear main seal with the block and also support the bolts which tighten the rear of the pan.

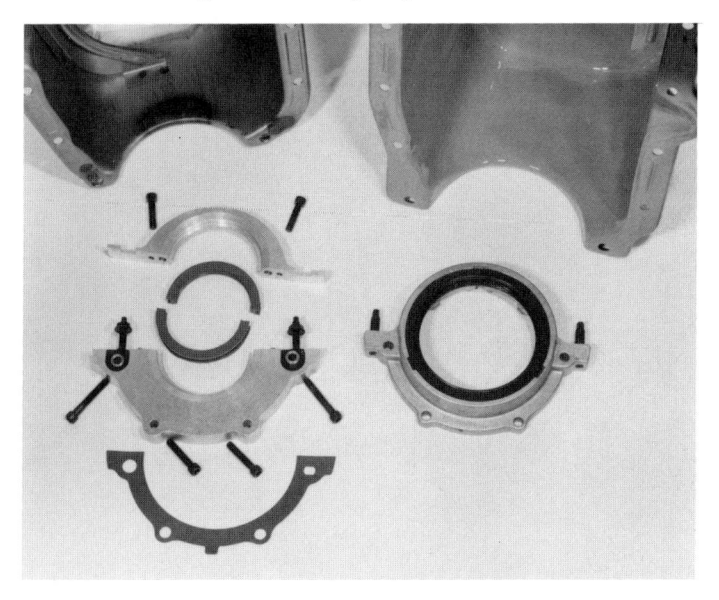

The rear main seal is installed last, after the crank and mains have been fitted. Bottom cap ties in to the upper one with two bolts. Rear pan studs also clamp the assembly together.

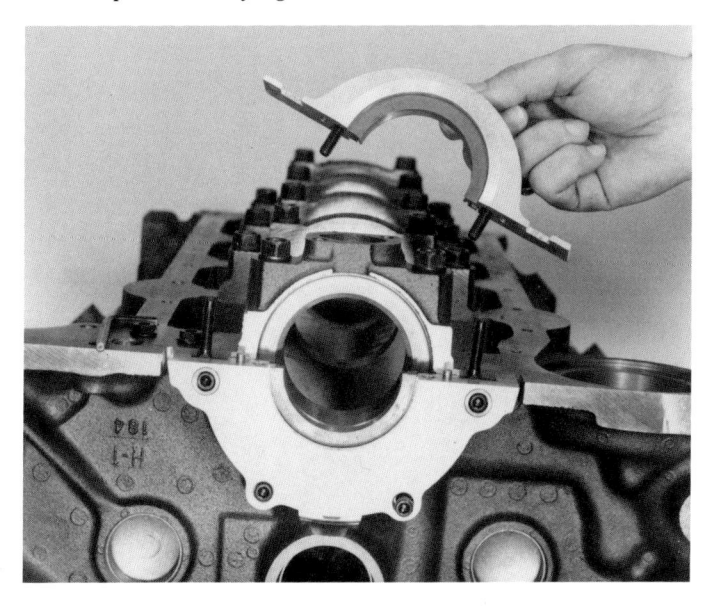

CHEVY POWER PARTS • 9

The Scoggin-Dickey rear main seal mates a new block to the old pan and crank. Oil drainback occurs between the rear main and the crank flange.

The aluminum block is delivered with splayed caps and pressed in centrifugally cast sleeves.

at the 58 cc combustion chambers gains fast burn and high efficiency. Increased velocity past the smaller intakes is also part of the picture. The fully assembled Corvette head, ready to install, is also less expensive than a matching Bow Tie. It is shipped with the GM silicon bronze 1.270 valve springs.

The one thing you will need with the Corvette head is a valve cover retained by four center studs. This, like the one piece seal oil pans, is a major leak eliminator. It uses a one piece neoprene gasket which really works.

The old style crank is counterweighted at the rear and also carries extra metal around the bolts. The new crank for the one piece seal requires a counterweighted flex plate, different from the older flex plate.

All coolant passages receive screw in plugs with O-rings and a rear cam bearing plug is replaced by a bolt on aluminum cover.

10 • CHEVY POWER PARTS

Yes, there are late model aftermarket valve covers which have the required clearance for roller rockers. Keep in mind that these heads have a taller rocker cover rail and this keeps the gasket from being flooded with oil.

VALVE TRAIN

GM rocker studs are available in two sizes, the 3/8 stud (3973416) and the 7/16 stud (3971912). Both screw into the same size hole in the head. The larger one is much stronger, but you then need roller rockers instead of the stock stamped steel bathtubs. The hardened guide plates are available for 492 and Bow Tie use (3973418).

GM made several interrelated changes on later heads, causing people who were used to the old format, to make mistakes on the new ones. The one thing you can't do is install old style rockers on new heads. Before doing any assembly, check the heads you are about to install. If the pushrods go through large round holes, the heads do not guide the pushrod. You can then use the late style bathtub rocker arms which carry a slot at the valve end. The slot pilots on the valve tip and provides rocker arm and pushrod alignment. The other way is to install guide plates and roller rockers.

If the parts department receives an anguished phone call about horrendous push rod and rocker wear, they know the answer right away: the voice at the other end of the line has tried to use the early rockers on late heads with predictable results. Either use guide plates and roller rockers or use no guide plates and Corvette rockers with a slot that

The Bow Tie block 10185047 has the stronger caps marked 2482. Some Bow Tie blocks have plain gray iron caps which are numbered 3412.

Chevy aluminum block is a stout weight saving casting with excellent finish and a lot of top notch detail work. It comes with steel main bearing caps.

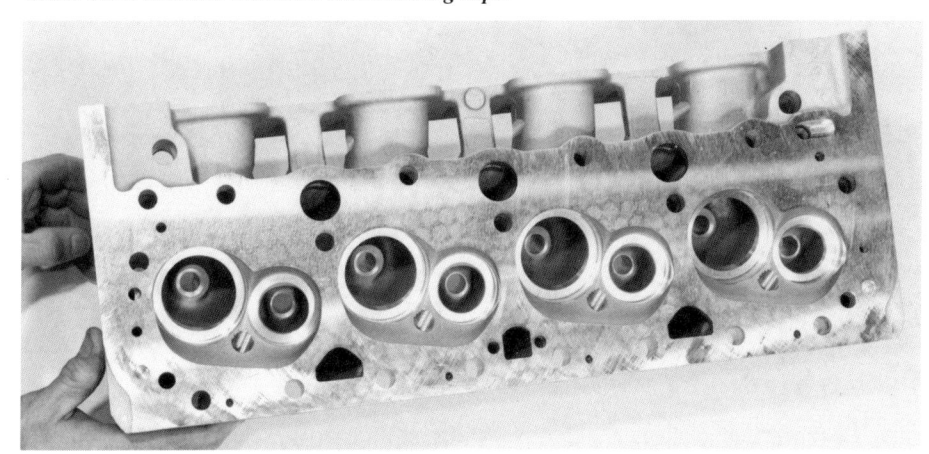

The canted valve head gains the same swirl and power in all cylinders. Install the seats of your choice on the intakes and exhausts.

The Bow Tie exhaust runners are the largest, for all out racing. The 492 has an intermediate exhaust port size and the Corvette aluminum head delivers max port velocity

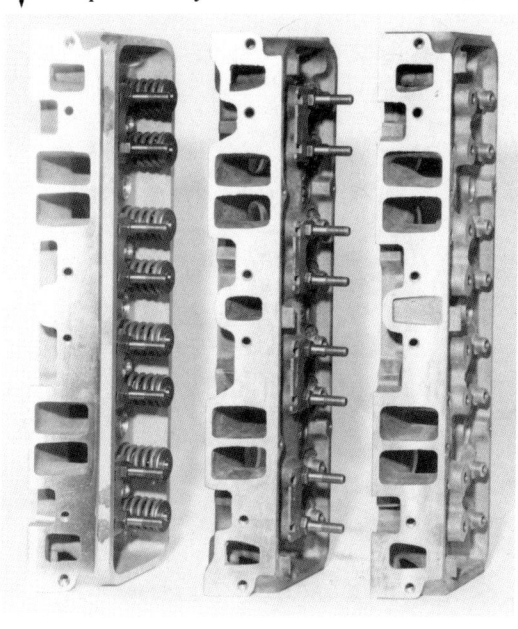

pilots on the valve stem tip. If you have late style net setting studs with a shoulder and no adjustment, simply convert back to conventional studs.

GIANT CHOICE

There is a vast choice of Chevy parts and the engineers at Motor-sports Technology are always developing new items. If you carefully select the right parts man, he will be right on top of the changes and what is best for your project. Having the right parts source is the secret weapon at many teams.

CHEVY POWER PARTS • 11

CRANKS STROKERS DAMPERS

Cranks can range from a cast street version good for 6000 rpm to a forged steel one shown here, or a billet steel one machined from solid forged billets.

You can improve the rod bearing journal oil gallery exit with a stone or a cartridge roll. Do not widen the chamfer excessively—it increases oil pressure loss.

There is no free lunch and when it comes to cranks, you get what you pay for—which is often more than good enough. We'll start with a practical and readily available cast iron crank, the kind you can find in any 350. It will work just fine under 6000 rpm although its real life range is under 5000 rpm.

It is not a race crank but it will run just fine in a budget street machine, longer than you plan to keep the car as long as you don't switch to 4.11 gears and play racer. The cast iron surface probably holds oil a little better steel and is easier on the bearings. Properly polished, the cast crank will last, as long as you don't overwork it with a blower or nitrous. When it develops enough cracks it splits.

A steel crank is essential from 6500 rpm and up, and useful at 5500 when the engine makes power and you hammer on it. Cranks are exposed to many vibrations and stresses that continuously vary, so crack resistance of a steel crank is very important. With steel it just happens at higher loads and rpm, and after more miles.

Telling them apart is easy: the steel forging has a smoother texture than a casting and a wider parting line. If you tap the crank with a hammer, the steel one rings with a higher pitch than one made of cast iron.

Chevrolet makes forged steel cranks for trucks without surface heat-treat. They also make heat-treated cranks for older Corvettes—they are harder, which improves bearing life and adds crack resistance. A good crank grinder generally has a heat treat source, but failures caused by excess crank twisting are alarmingly common. There are many superior billet steel cranks where the crank designer is not limited by an available forging and can add extra material in the right places for strength and balancing. There are also special forged cranks for strokers.

STROKERS

The least expensive stroker crank is made out of an existing crank by offset grinding. Here the journal is ground smaller, and also offcenter, to gain extra stroke. Removing more material from the inside than from the outside moves the center of the pin and increases the stroke. One convenient application is a 350 crank reground to 327 journal sizes which brings the stroke from 3.48 to 3.562 and delivers 366 cubic inches. Another move is to regrind a 400 crank and install it in a 350 block. Now you get 377 cubic inches from a stock 4.00 bore and 383 CID from a 4.030 overbore. The main bearings are cut down from 2.649 to 2.449. You end up with the closest thing to a 400 engine without having to contend with heat from siamesed bores.

> *In compiling the material for this section, we received an enormous amount of help from Dick Titsworth of Seaport Automotive in Toledo, OH and Bob Gillelan of Moldex Tools in Dearborn Heights, MI. Both have raced extensively and have ground cranks for decades. We were also assisted by Bob Patzold of Prototype, in Hainesville, IL, whose welding equipment is in these photos.*

12 • CRANKS - STROKERS - DAMPERS

Best crank welding is done with a submerged arc—the flux material gasifies and protects the weld area from oxides and inclusions.

Some of the weld buildup is directed at the radius and cheek, the rest at the journal.

Remove all sharp edges and debur the crank prior to balancing.

Chevrolet Engineering's Raceshop offers steel rough forgings with ample stroker material and wide arms that are popular with crank grinders. Some crank grinding shops apply a weld build up for repairs and stroking, using a wire feed under a submerged arc. Flux fed out of a hopper gasifies and forms a continuous shield around the weld area. The type of flux used does have an effect on the hardness of the buildup.

There is no great trick to building a stroker engine but here are some things to keep in mind. A higher stroke works best if you can combine it with a longer rod. This keeps the rod angles from becoming excessive and reduces the side thrust pressures on the cylinder and piston. Longer rods and strokes push the piston further up out of the bore so now you need a piston with a much lower compression height—less height from the center of the wrist pin to the top of the piston. The outer limit is when the top or the rod is too close to the piston top and when the oil ring conflicts with the wrist pin bore. A stroker also calls for a shorter piston skirt which won't run into the counterweights. Here the long rod helps by gaining skirt length.

A stroker is better in many ways than a block bored to the max. It leaves more cylinder wall thickness and produces good torque. However, a stroker is more rpm limited.

With a stroker you need to check for interference between the bottom of the piston and the crank, at the sides of the block, at the bottom of the cylinder barrels, and between the rod and the cam. With larger strokers you will need a cam ground on a smaller base circle. When you order rods, pistons, and cams for a stroker your supplier needs to know ahead of time what you are building.

INSPECTION TIME

Any crank can look good, but still have cracks. As with potato chips, you can hardly ever have only one crack, and most of them occur in the radius areas at the main and rod journals. Make it a practice to Magnaflux all cranks, new or old. The cracks are generally tight, too fine to see with the naked eye, and when the cracks are large enough to see, the crank is usually junk. Vibrations and cycling invariably spread a crack into full breakage.

Really expensive cranks can be repaired by grinding out shallow cracks and rewelding. With cheaper cast or steel cranks, repairs usually costs more than replacement.

The one thing all cranks need, new or used, is a Magnaflux inspection: most cracks occur in the radius areas where the journal joins the cheek.

If you see a crank with a blued, scored journal you know that it has seen heat. Reginding a crank that turned blue becomes a matter of judgment: talk it over with the crank grinder. Also, get to the root of the problem and find out what caused the failure. That means trouble shooting everything from the pump and its relief valve to the filter and the oil galleries. Also, cut open the oil filter to see what debris traveled through the engine. Check the cam and lifters for score marks and wear. A burned rod usually indicates oil starvation. A failed main is usually due to some overload during operation.

A scored crank can be reground to a stock stroke undersize or can serve as a core for an offset ground stroker. A regrind to 0.010 to 0.020 under is safe and normal. For a grocery getter, 0.040 is the limit. You can get 0.001 and 0.002 undersize bearings for crank fitting and it is acceptable to mix and match bearing shells to gain the right clearance.

Reground cranks are not all of the same quality. Check that the journal meets the radius smoothly without a step. Shiny is not enough: mike each journal in the center, at the ends, and then repeat at 90 degrees to the first measurements. You should hand polish all reground crank journals with scuffed up #400 emery cloth.

OILING

The block oil galleys supply oil to the crank drillings at the mains and from there the oil travels to the rod journals. None of this is as pat as it sounds. All of the rod journal oil supply is delivered by the grooves in the top mains. You are likely to get more oil to the journals when the block and crank galleries face head on, and slightly less when the oil has to travel through the groove.

As the crank turns, it tends to entrain oil with it, in the direction of crank rota-

CRANKS - STROKERS - DAMPERS • 13

A forged crank will have a fine gray texture and will show wide forging trim lines rather than thin foundry parting lines. This one from Diamond Racing has been extensively prepped.

tion. As a result, the front part of the groove does not supply oil as well as the rear part. Using grooved bearing shells in the main bearing caps may improve the oiling but is directly detrimental to the lower bearing's load capacity—very significantly so.

You can improve things with cross drilled mains. On the number 1-3-5 mains single cross drillings are sufficient, while number 2 and 4 mains need a more complicated T-shaped drilling because the oil galleries are 90 degrees apart. You can insert welding rods through crank oil galleries to see how they enter and exit. Changing crank drillings is not a do-it-yourself project and your crank grinder will know where and how to improve them.

CHAMFERS AND LEAD GROOVES

An oil gallery without an entry or exit radius will restrict oil flow. Generally, a 0.025 to 0.030 radius helps turn the oil into the crank from the main bearing side. This radius should not be widened toward the sides at the mains because that would tend to create a leakage path toward the edge of the bearing. The exit part of the gallery at the rod journal can be a little wider because you are trying to spread the oil through the width of the bearing.

It is common to add a lead-in groove from the main bearing oil drilling, in the direction of crank rotation. Now, the oil feed gallery can pick up oil a little earlier over a longer period of time, which helps the rod bearings. Do check the location of the oil galleries at each rod journal; even with factory cranks they are sometimes not centered with the bearings. There is little that you can do about it except change cranks.

CRANK RUNOUT

You carefully stand up the crank, but someone walks by and bumps into it. Being very polite, he picks it up without saying anything, except that now the crank has a runout. You cannot spin a bent crank with your finger tips as easily as a straight one and you are building the best. Some crank grinders suggest between 0.0005 and 0.001. Others feel that up to 0.0025 is tolerable.

Either set the crank on a pair of Vee blocks or place it on two bearing shells in the block and use a dial indicator to read the runout. Beyond 0.005 inch it may be not just bent but also cracked. To add a little confusion a crank does flex all the time during operation under firing loads. Cranks have been deliberately run on durability with 0.005 runout and lived, but it certainly isn't the accepted way of building an engine.

THRUST SURFACE

The rear thrust face accepts thrust from the clutch or the torque converter (Don't laugh, torque converters do expand lengthwise). The higher the spring rating of the clutch, the more likely the damage. To offset the effects of high clutch spring pressures, Dick Titsworth has installed a girdle at the front of some of his engines and mounted a throwout bearing on it. Now, when the crank moves forward, it contacts the throwout bearing first, which prevents thrust surface damage. Crank end thrust is normally limited to 0.005 inch to protect the timing chain and to provide enough oiling at the thrust surfaces.

CRANK DAMPERS

A 350 crank is strong, and the one in the big block is even more massive. You would think that it is infinitely stiff, but it isn't. In effect, it acts like a torsion bar and twists back and forth as it continues to turn. The total amount of twist or amplitude can reach up to one degree under severe torque loading. Each power pulse twangs on it at a different point. While the inertia of the vehicle is acting at the flywheel flange, the other end is free to wind and unwind.

Superimposed on these torsional vibrations are a number of bending moments due to unbalance forces and they go up as the square of the speed. Now, instead of considering the crank as infinitely rigid you can look at it more as a live spring. It has its own natural frequency, just like a drum or a guitar string and it tends to resonate or sound off at that frequency and fractions or multiples of it. When you can feel the sound of the band through the glass at your table, you begin to understand that vibrations are powerful stuff. When engineers speak of a first, fourth, or sixth

In the Fluidampr the ring is on the inside, totally contained in the housing and the cover is laser beam welded on the ID and OD. The two holes in the cover are used to fill the assembly with high viscosity silicone and are then factory welded shut.

order of vibration—they simply refer to fractions of the natural frequency.

Say a crank has a natural frequency of 20800 hertz or cycles per second–the shake rate at which it resonates. A quarter (1/4) of that natural frequency corresponds to 5200 rpm. Don't be surprised that this fourth harmonic is one of the highest shake modes for a V8 crank. Once the natural frequency and the impulses are in phase with each other the amplitude or twist increase substantially and can cause crank failure.

You have taken out rattles by placing your hand on the glass or on the glove compartment door. By the same token, a crank damper mounted at the nose of the crank tends to control the amplitude of vibrations occurring at certain rpm.

A stock crank damper has a hub keyed to the crank and an outer inertia

14 • CRANKS - STROKERS - DAMPERS

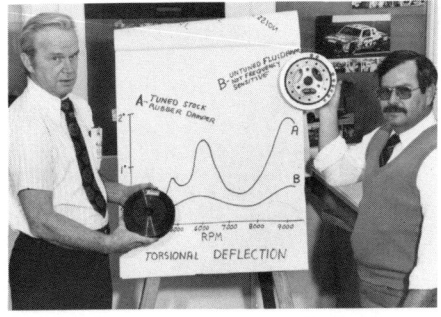

A running crank constantly twists ahead of the direction of rotation and then lags behind it. It is up to the damper to quiet the oscillations. Rubber dampers do not like ultra high rpm.

ring. Between those two is a layer of special rubber. As the crank nose twists ahead of the rotation, the inertia ring tries to continue running at the original speed and hangs back. Then, as the crank twists the opposite way and its nose lags behind the direction of rotation the ring is faster than the motion and again damps out the crank vibrations. Keep in mind that because of its mass and inertia, the ring doesn't want to know about vibrations—it just wants to travel at a steady speed.

The stock street damper has seen a lot of racing but it has its limits because the cast iron ring does not like high rpm. It has been banned from many competitive events because too many of them have blown apart. Mandated right now is a steel ring—this steel leaves us with a rubber insert which also does not like high rpm. Eventually, vibrations, oil, and heat get the best of it and you see it tired and checkered. Smart racers mark a line or use punch marks to locate the ring in relation to the hub. When they see the marks move, they know that first of all the timing marks are inaccurate and secondly that the rubber is tired and ready to give up.

The smartest replacement damper to have come along is made by Vibratech,® (formerly Houdaille) in Buffalo, NY under the tradename of Fluidampr.® A hub, fully machined inside and out, forms the back half of a ring-shaped housing for the inertia ring. This hub is keyed to the crank. The ring is first machined, then coated with a black nylon, after which the nylon is machined for a precision fit inside the hub and housing.

When the Fluidampr is manufactured, the ring is inserted in the hub. Next, a cover is pressed in, so it seats at machined pilot areas of the hub and is laser beam welded at the OD and ID, fully enclosing the ring. There are no seals to wear out and no place for dirt to get in. Next, a filling machine applies a high vacuum to one small hole in the front cover while the other small hole serves as a supply spot for some amazingly viscous silicone fluid. A film of silicone hugs the inside of the housing and the outside of the inertia ring, tending to hold the two together.

With the engine running at maximum load, the Fluidampr sees around a half a degree amplitude, far less than with rubber type dampers. By changing the viscosity of the silicone and the clearances involved inside the damper, they can fine tune the unit to a particular application. The ring is not connected to the housing except by the silicone and rotates within the housing without, in any way, affecting the timing marks engraved in the outer housing. It is the viscous shear forces in the silicone that provide the damping action, not any particular position of the ring and the outer housing. With a ring held by rubber, the ring must return to its initial position, not so with the Fluidampr.

John Lingenfelter tested a number of different damper designs and found 5-15 horsepower improvements from quieting down the crank and also tells us that the bearings look better. Engine builder Ron Hutter also confirms this.

We get the feeling that the crank not only oscillates in torsion but also forms a lengthwise wave shape—the kind of wave you see in a string that you have just twanged. In this respect, it will have nodes, just like the places where a twanged string stands still or has areas of high amplitude. When you quiet down the torsionals, you may also keep the crank out of the lengthwise vibrations-there is less bearing drag and power gains to show for it.

Many sprint car engine owners are at the opposite end of the spectrum and use no damper because the front of the engine is encumbered with fuel pumps and power steering pumps. They also break a lot of cranks. Bill Howell points out that a sprint car engine has no flywheel to act as a base against which the crank can twist. When you try to run a rubber inertia ring damper on a sprint car or midget, it will fling off the inertia ring in less than five laps. In spite of all this, sprint cars built with good cranks, a reasonable stroke length and arm width have only modest crank breakage problems.

Fluidamprs are available neutral balanced or with external counterweighting. ➝

The viscosity of the silicone filler in the Fluidampr increases as the crank torsions. Bo Laws and John Lingenfelter both tell us that bearing life improves.

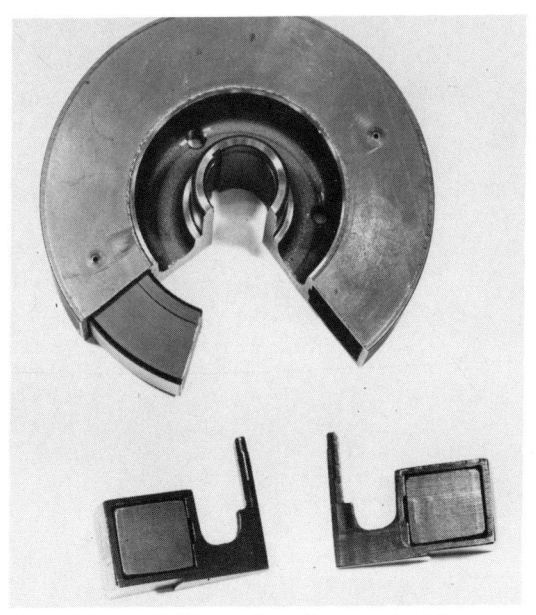

The ring is fully machined, coated with a black material to friction-proof it; and the coating is also precision machined.

CRANKS - STROKERS - DAMPERS • 15

CONNECTING RODS

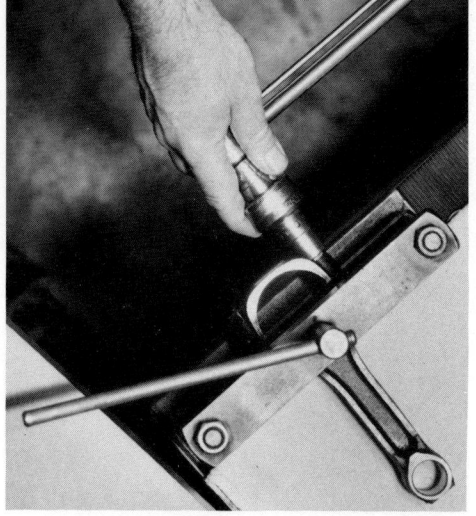

The rod is held in a vise to keep it from being twisted as the bolts are loosened up or torqued. The parting face is reconditioned and new bolts pressed in.

People work on carburetors, heads, cams, almost anything that makes power. Rods? Almost anything will do. "After all, if I had the extra bucks I'd plow them into big valves and make some real power..." The rod gets even in its own small way—it just breaks and takes everything in its path with it. Of course, the crank turned blue so it must have been a bad bearing or a bad crank. Point is that the rods sees high stresses and the strange part of it is that the power is not what breaks the rod.

In fact, the power stroke represents the least stress on the rod, compared to

Most street piston and rod assemblies are a press fit. To remove the pin it must be pressed out, and the piston needs support to prevent damage during this operation.

the transition from exhaust to intake. The full weight of the rod and piston is accelerated from BDC on up. As the piston reaches the top of its travel something must make it stop and reverse direction. The stretching loads on the bottom end of the rod at that moment are very high and the rod bearing also wipes the bottom of the crank. During compression and power, the pressure on the piston helps oppose the forces trying to send the piston through the head and so the load on the rod is actually reduced. High rpm coasting with the throttle off will help take out a rod much faster than a power run up the previous grade.

ROD RECONDITIONING

The big end of a rod is bored round when it is new but it doesn't necessarily stay that way. A stretching action tries to ovalize the big end, pulling it in along the parting face. Every rod is worked on that way each time the piston passes TDC. A good rod designer adds material at the parting face, bigger bolts, provides stiffening pads at the corners, better materials, and when you pull the rod out of the engine Magnafluxing and dial bore gauges tell the tale. Rod bolts are not the place to save money—get the good ones that can do the job.

With a street engine, stock gears, and moderate power, minor reconditioning will do. For a high powered street machine or a bracket racer, the rod bolts should be replaced with high grade part or spring for good high grade rods.

We shot photos of rod rebuilding at Ralph Thorne's South DeKalb Machine Shop in Decatur, GA. Each rod is degreased, Magnafluxed to check it for cracks, and also checked for the rod and cap belonging together. Also, the rod bolts are pressed out so as not to be in the way of regrinding the parting faces.

The big end of a rod becomes elongated—stretched—and is larger along the length of the rod than across the parting face. Before we can make the bottom end round, some material will have to be removed from the parting face at the rod and cap. This achieves two purposes—flattening the surfaces and shrinking the hole enough so that there is some metal for the hone to bite into.

The rod is clamped into a special fixture which automatically squares it against a flat fixed horizontal plate and against the sides of the clamping device. The rod and clamp are then swung past a spinning cup shaped grinding wheel and resurfaced. The same procedure is repeated for the cap.

The next move is to press in the new rod bolts and to then clamp the rod in a special vise to keep it from twisting while the bolts are torqued. Now the big end up the rod is ready to be honed out round. Ralph uses the stroking device on his Sunnen hone to get a correct cross hatch pattern for bearing retention. The rod bore is always slightly smaller than the bearing to ensure sufficient "crush" or interference fit. If you swap rod bolts, do rehone the bottom end. When you press in the new bolts, they expand and deform the metal at the bearing bore.

Once the big end of the rod is completed, the small end is checked for size and if need be honed out. With a pressed pin you need to retain a 0.001 interference fit. For full floating pins, rebush the rods as part of the rebuild.

ROD LENGTH

When the crank is at TDC the rod is directly in line with the axis of the cylinder. As the crank turns, the rod angle with the cylinder keeps increasing and so do the side loads. It works just like changing the angle of a ladder against the wall. Naturally, the longer the rod, the smaller the angle of the rod and the lower the side thrust. In addition, there will be timing changes that cause the piston to dwell a little longer near TDC, than at the bottom of the stroke.

The small and big ends are now honed to size. The bearings are slightly larger than the big end and the crush retains them.

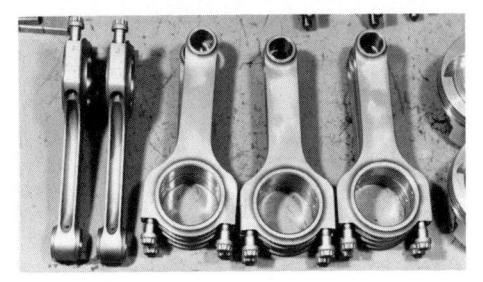

It is all a matter of degree—Fred Carrillo may spend more on one rod bolt than the cost of a stock replacement rod. The rods are milled out along the side of the beam for lightness and have rounded shoulders so a stroker can clear a cam. The rods are shot peened and polished with no balance pads at either end, to save grams.

A rod boring fixture is used at Prototype to have the center to center distance identical. The top and bottom ends have been reconditioned and the new bushing has been pressed in at the small end.

When building a durable Winston Cup style engine, most racers try to cram the longest possible rod into the block, going from 5.7 to 6 inches or more. You soon run out of room because the piston compression height can only be squeezed so far at which time the top of the rod nearly touches the piston and you see oil rings running through buttons in wrist pin, or two ring pistons.

One thing is certain, a long rod weighs more. Generally, a long rod engine will pick up some top end horsepower while the short rod engine is credited with an increase in mid-range torque. With a short rod engine, the piston moves away from the head with more initial velocity, gets a stronger pulse started and this can be turned into a breathing advantage. As you can see, there are enough pros and cons about rod length listed here to get an all night argument off to a good start.

As the stroke increases, the stroke to rod ratio changes unfavorably and now you are looking to make room for not only the larger crank but also the longest rod that can fit. You need one, if for no other reason to get the bottom of the piston away from the crank. Once you have satisfied that requirement the long rod and piston need a taller block deck as in the truck type big block.

With a small block you have a choice of an infinite amount of different rods and one of the choices you can make is between 2.0 and 2.1 inch rod bearings. In a drag engine the smaller bearing offers slightly lower linear speeds and the larger one allows a more durable crank pin. In a street application 2.1 pins are your best bet. For serious power, consider the Bowtie rods first. They are available in 5.7 and 6.0 inch lengths and have ample extra metal in the right places. If you are going to spend more money, there are top rods made by Carrillo, Crower, Oliver, Lentz, and a number of others.

You may decide to polish the rod body and reshot peen it. Also a rod can benefit from being lightened. It makes sense to do so, but first we would like to leave you with a couple of thoughts. Any rod fails rather instantly from surface cracks. Excess grinding heat will create cracks. The idea is to keep the heat at a minimum. Any grinding or polishing rods should be lengthwise with the rod rather than crossways—you don't want cross cracks at the highest stress area. If you are going to do some grinding you should also spend the time to sand out and polish the rod surfaces to eliminate stress risers. A little metal removal from a sharp corner creates a radius and substantially strengthens the rod. This holds particularly true at inside bolt corners.

Lighter rods do help reduce bearing loads and the easiest place to remove metal is at the balance pads, on either end of the rod. With a jig based on a wrist pin sized pivot, the top end of the rod can be radiused on center without relying on hand work in front of a grinder. If in doubt, leave a little extra metal, in place because the top end of the rod is a fragile area. In the case of a stroker, you will generally need some grinding at rod shoulders. Have a little conference with your machine shop or your balancer before you dive in there with a grinder.

Boring a rod on a fixture ensures that the centerlines of the small and big ends parallel, and that all rods will be at the same center to center length. Uneven one-sided piston wear tells you if a rod is bent, pulling to one side, is twisted or imparts a cork screw motion.

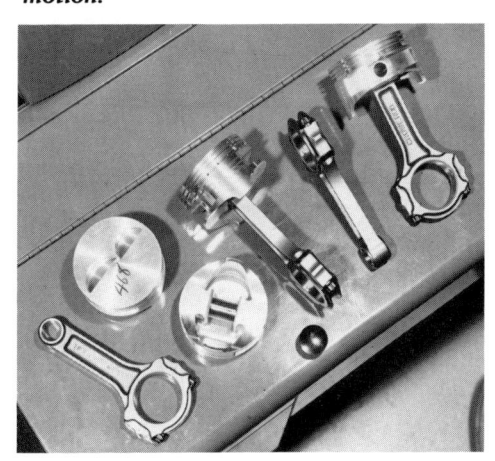

The wide parting face of the Crower rod resists the hooping action as the rod is being stretched by the changes in piston and rod direction at TDC.

Rods bought at an auction are very useful as paperweights, boat anchors, ballast, but they really don't belong in an engine. If you buy used rods at least know the guy you bought them from.

When you install reconditioned rods, break the corners left at the parting face, so you don't scrape the back of the bearing shells as they are installed. Draw a file along the corners, but don't let the file even touch any part of the rod bolts. Tape for protection.

CONNECTING RODS • 17

PISTONS

A forged piston begins as a slug of aluminum alloy which is squeezed between an outside and an inside die, forced up from the crown to form the skirt and the two pin bosses. If the piston is designed right, and most of them are, the forging improves grain flow and provides substantial strength. Forged pistons measured in a lab at room temperature are definitely stronger than castings. When an aluminum piston reaches operating temperature, a good part of that strength is lost, regardless of which piston you get. It comes with the territory, and varies with the alloy used by the manufacturer.

A forged piston will be stronger if it does not have a straight slot behind the oil ring groove. Drilled oil return holes are preferred, and even then, they should not be right at the pin boss center or at the center of the skirt, which are high stress areas.

There is a wide variety of skirt designs ranging from narrow to wide slipper skirts. Generally, a skirt that offers a wide support area against the cylinder runs at lower unit loads, does not deform the cylinder walls as much, lives longer with less drag, and is less likely to score. Skirts with a support hoop near the bottom seem to retain good oil control and piston stability. Pistons with forged side reliefs have excellent weight and strength control. A skirt needs a certain amount of flexibility and should conform to the cylinder.

When you see an even wear or contact pattern on the skirt that has run, you know the designer did his homework. A piston is purposely made with the cam contour and a skirt profile at room temperature so that it can be round and "straight" at operating temperatures. The top of the piston runs hottest, so its diameter is purposely made smaller. It also gets a a slight taper at the ring land area. This is done so that when the piston rocks within the cylinder bore, the land area will be parallel with the bore and maintain even clearance. The skirt is smallest at the top but increases in size toward the bottom, in a profile decided by the temperature gradient to compensate for expansion. The trick is to have the skirt flat against the cylinder, with the piston at operating temperature. The piston also grows more across the pin towers than it does across the thrust surface and this calls for a cam contour.

One of the critical areas is the tie-in between the skirt and the pin bosses. When the transition is abrupt, cracking problems take place. Curved sections between the pin boss and the skirt, as well as the radiused areas between the piston crown and the pin bosses are all related to durability and avoiding sudden cross section changes.

HYPEREUTECTIC

Forged pistons are a must above 6000 rpm. In a range from 5500 on down there are several alternative pistons that have proven extremely effective in terms of quiet operation, heat resistance and durability. One interesting engine building option is offered by Speed-Pro's Hypereutectic pistons, which are made out of an alloy with a high percentage of silicon (16 % instead of 9 %). The silicon we speak of is an element which has nothing to do with the sealer silicone. You will find it in sand by the sea and in your your windshield glass. Molten aluminum will dissolve as much as nine per cent silicon, making the aluminum harder and more stable, once it cools down. Silicon aluminum machines more like cast iron

If you use Spiro-Locs and want to check the spacing between the grooves this inside mike is the one to use. Just measuring pin end play is inadequate because the Spiro-Loc tends to expand and the spring action gives a false reading.

Ring lands need to be absolutely flat so that the ring can seal against the piston as well as against the cylinder. Piston turntable rotates slowly as electronic gauges measure it.

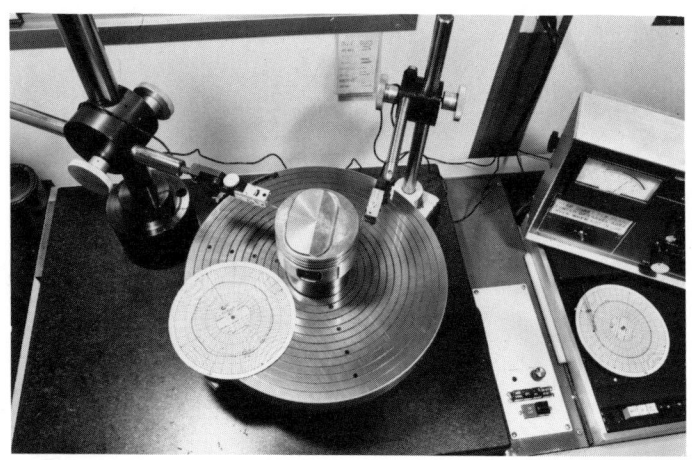

Cutaways through the forged and Hypereutectic Speed-Pro pistons show some interesting differences—and each has a specific purpose. The forged piston is great for racing and for top power output—note the straight pin bosses and ample radius between the pin boss and the skirt as well as across the dome. Hypereutectic piston is lighter, has very even sections, and less mass in the pin boss areas.

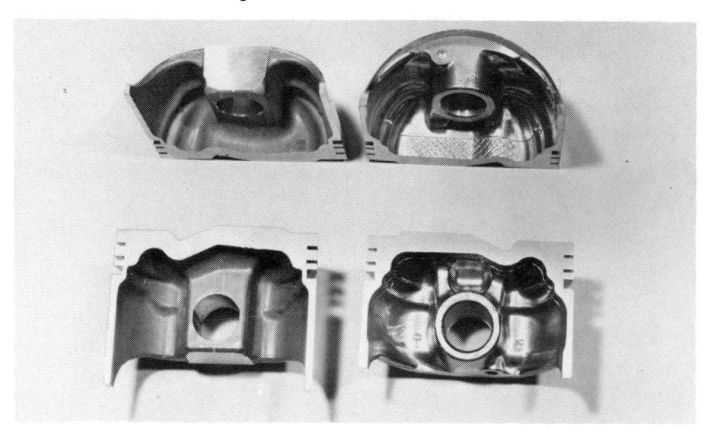

18 • PISTONS - WRIST PINS

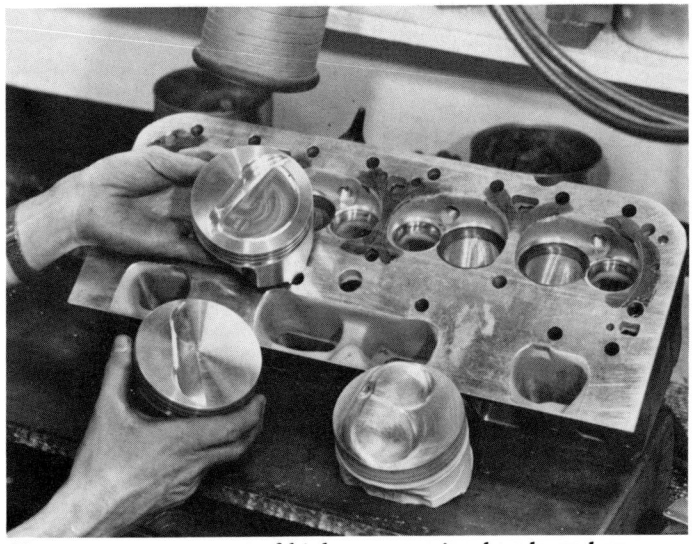

You can preserve a set of high compression heads and run them with either pop up pistons or with dished pistons to form a 9:1 engine.

You have a choice of piston top shapes from dish to flat top and domes shaped to match different combustion chambers.

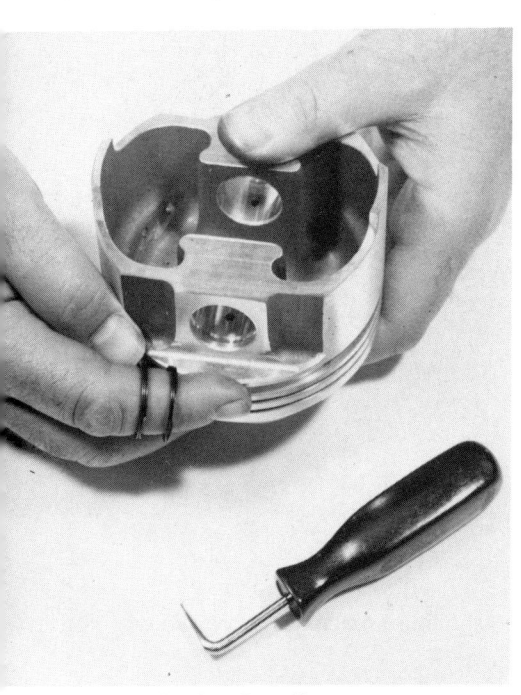

Diamond's Jim Cavallaro remanufactured a right angle screwdriver with a pointy tip to install the Spiro Locs. Use an awl to remove them.

than aluminum. It also expands less, can be run at tighter clearances within the cylinder bores, hence a quieter engine with better ring control because the piston doesn't rock as much. Hypereutectic is an advanced alloy with 16 per cent of silicon. In fact, its name means that more alloying silicon has been added than the molten aluminum will dissolve. The trick in that alloy is to have fine small grained silicon particles dispersed through the alloy. To explain what happens let's side track for a moment.

We normally drink it black, but on this occasion we are going to dump a spoonful of sugar into our coffee. It dissolves, so we put in a little more. After a while it does not want to dissolve anymore, so the sugar goes to the bottom of the cup—the coffee is saturated—that's about the level of a nine percent silicon in aluminum. However, let's try something different—let's stir the coffee while it is boiling hot and able to dissolve a lot of sugar, put it in our super instafreezer and all the little sugar particles are frozen in space. Now you are coming closer to the new Hypereutectic alloys. They carry more silicon than the aluminum is supposed to, all finely dispersed throughout the metal. The machined Hypereutectic piston glides on those ultra hard silicon particles. They are cast by a special process with a fast cool down and one

The interrupted surface finish prevents scoring and metal pickup. We rubbed ink into the grooves to make them stand out. Incidentally, the profiled grooves are individual—not a thread finish.

of the advantages is excellent control over the thickness of all the sections. As a result, they are unusually light for their strength level and easy on the bearings—also rev up quickly.

Zollner Pistons and Speed-Pro came up with a vast array of those performance Hypereutectic pistons; Chevrolet uses some of them and so do many hot running marine engines. They have excellent high temperature resistance, exceeding that of forgings. We have used some on street engines and they run quietly in a 10:1 motor with substantially less clearance than a forged piston, and at a very reasonable cost. John Lingenfelter has built some strong engines with both forgings and Hypereutectics and comments that you can get shapes and weight savings out of the Hypereutectics that make it very competitive in a moderate rpm engine.

RING GROOVES

The most popular ring combination calls for 1/16 compression and 3/16 oil rings. They are the easiest set to find on short notice. You can order pistons with tighter grooves and hand lap the compression rings to just fit the grooves. The 0.043 ring is lighter, has less inertia, and reduces friction by having 2/3 the width of a 1/16 ring. Standard street 5/64 rings are just wide enough for good street use. They happen to be 25 per cent wider than a 1/16 ring and that extra thickness definitely has an effect.

When space is at a premium and the wrist pin is moved up to make room for a stroker or a longer rod, the first move is thinner ring lands, which weakens

This Diamond Racing machine follows a pattern and produces the 3-D skirt contour. There is a cam shape built into the skirt as you look at it from on top and also a ski-shaped profile, as seen from the side.

them. You can gain a little space with a thinner top ring. For drag racing you gain still more space with two ring combinations or by having the oil ring run in support grooves machined in plugs at the sides of the wrist pin.

WRIST PINS

Everyone wants to lighten the reciprocating assembly. They get mesmerized by the appeal of a few grams and switch to thin pins or tapered pins. Diamond Racing Products' Jim Cavallaro tells us that pins are the wrong place to save weight. Your average four inch piston has a shade over 12 square inches of area and at the peak of combustion each one of those square inches sees 1200 psi or better. What do you think happens to a skinny wrist pin? If your magic high speed camera could film it you would see that instead of being round, it just momentarily squashes and becomes 0.002 or 0.003 inch wider and lower. That eats up the slight piston to pin clearance and gives the pin quite a grip on the piston which now galls, picks up metal, and destructs. The pin springs back and looks good; the piston looks terrible, so the piston gets blamed when what you really needed was a stout pin that does not deflect. Winston Cup racers go for 2.950 pins while short track racers prefer shorter 2.750 pins. A 0.927 pin diameter is standard for small blocks while big blocks use 0.990 pins. A short track motor can run a 0.145 wall pin, Winston Cup engines receive a 0.165 wall pin, and turbo engines live longer with 0.185 wall pins.

VALVE POCKETS

Establishing the correct valve pocket depth is very much part of prefitting an engine. You need a minimum of 0.100 to 0.125 clearance between the piston and the valve in a street engine which will be shared by you with people who don't know as much about engines as you do—that is insurance clearance. When you build a drag engine with 12 or 13:1 compression in a small cube block all the rules change. However, you know ahead of time that missing a shift will cost a set of valves, as minimum fee, and you play with the tightest clearances one can get away with.

For starters preassemble the engine with the crank, one rod and piston, no rings, the cam you are going to use, degreed in. Use one or two valves at a time in the head, installed with very light checking springs. Add a dial indicator at the valve and a degree wheel. As you turn the crank, keep pushing down on the valve to see how much room is left between it and the piston.

A piston chases the exhaust valve on the up stroke and is chased by the intake valve on the down stroke. You will generally pick up the smallest exhaust to piston clearance at 8–10 degrees BTDC (before top dead center) and the least intake clearance in the same range ATDC with the piston on the way down from overlap.

Naturally, all rules change if the cam is advanced or retarded. If you plan to work at the drag strip with different cam timings the valve pockets need more depth for a wider range of conditions. Make up a little chart and record the depth of the valve pockets needed.

The higher the compression the more you start chasing around in circles, because cutting valve pockets takes away from the compression you are trying to build.

MACHINING VALVE POCKETS

To machine an accurate valve pocket you first need to locate it in the right place on the piston by marking a center. This is done by using a long transfer punch of the same diameter as a valve, and with an accurate center at the tip. The transfer point is inserted into the guide after the piston has been brought up to the minimum piston to valve clearance, as shown in the previous section. A little Dykem blue on the piston makes the mark stand out. The transfer punch can be bought, or fabricated from a used valve.

A machine shop can easily put in valve pockets without your center mark: they assume where the center should be. Since they don't really know, without all the parts in place, they play it safe by cutting a little extra. When you give them a center, the

Pin height is measured from the center of the pin to the piston deck. With a longer rod you'll need a reduced pin height. Stroker motors also call for less pin height. Bottom of the skirt is cut to clear crank counterweights.

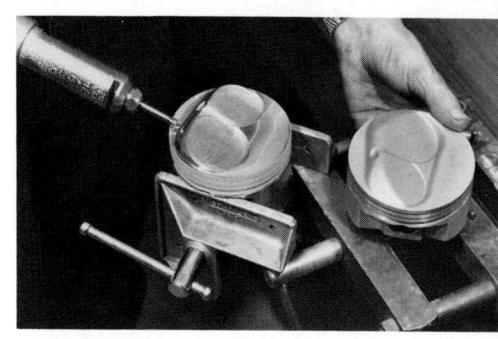

A little prep work eliminates machined edges and hot spots, gains a more professional look, and improves flame propagation.

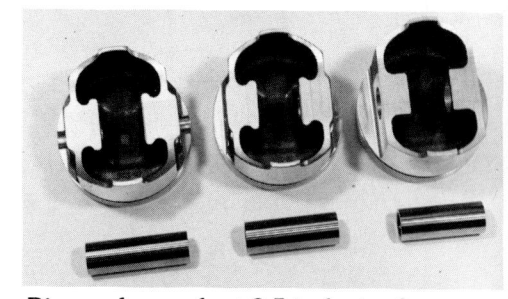

Diamond uses short 2.5 inch pins for 9:1 short track racing; the 2.750 pin length offers additional support area but most Winston Cup engines are built with 2.950 pins.

pocket is put in right on the money, where it should be and you do not lose cc needed for compression.

If you have access to a mill you can put in your own valve pockets. PMS makes a great piston vise which bolts to the mill. Unlike some vises which position the piston by ring grooves, this one holds the piston by the wrist pin which results in the more accurate location that corresponds to what you have in the engine. You actually use two pins, one in each pin boss and they locate on a pair of Vee blocks that form part of the vise. PMS also offers the necessary fly cutters to machine the pocket. Pays for itself in the first couple of engine jobs. Incidentally, holding the piston by the pin, the way it is located in the engine, is more accurate than holding it by ring grooves. When you do eight pistons in a row the ring grooves may vary while the pins are on the money.

PISTON DOME THICKNESS

A piston dome sees a lot of heat and pressure, so the more you cut the more likely the piston is to cave in. As a rule you need at need at least 0.200 inch thickness in the dome and valve pockets. You should use up some form of piston dome mike. Dick Titsworth at Seaport made his own from a dial caliper. PMS supplies one which is simple, light, and inexpensive and B-H-J makes a dial indicator stand with a pick up rod on the underside and an indicator on top. Whichever one you use, measure at right angles to the surface to get an idea of metal thickness and check valve pocket corners because they are the high stress points. If you machine a piston or have one done, make certain a radius is formed in the corners to avoid stress cracks in the pistons.

DOME PREP

In about 20 minutes or so per piston you can prep the top so that all corners are rounded, shaped, and finished. Flame travel does not like sharp edges and smooth contours will help more than saving that extra cc. You can smooth the piston with a carbide cutter, and cartridge rolls. Butch Elkins over at Diamond Racing Products showed us how to prevent metal buildup in the cutter grooves, by dipping the cutter into oil, grease, or wax. Just round off sharp piston edges without removing massive amounts of metal. Next, you can follow up with a cartridge roll to smooth out the entire top of the piston.

Whether you are using a carbide cutter or a cartridge roll, never stop moving the tool or you will create a dig mark. Practice on a scrap piston first.

BEAD BLASTING

Bead blasting will put a beautiful finish at the top of the piston. However, if you bead blast the entire piston, you have just created the world's most expensive oil pumper. Ring grooves cannot tolerate bead blasting and neither can the pin bores or the skirts. You

Locating the piston by the pin as on this PMS vise is more accurate than grabbing it by the ring grooves. Use the center punch mark on the piston to center the cutter.

should tape up the entire piston to protect it before bead blasting the top. If you do not need the cosmetics, keep pistons away from the bead blaster.

COMPRESSION DISTANCE

Bill Willett, Speed-Pro's piston engineer, gave us unlimited assistance with this entire piston section, and one of many interesting things he covered was compression distance, a term that helps you fit the right piston into the engine. That compression distance extends from the center of the wrist pin to the deck of the piston—the flat part. It does not include dome height.

With the piston at TDC and the crank pointing up, you add up half the crank stroke, the rod length, and piston's compression distance. The total cannot exceed the space from the center of the block to the top deck. If you switch to a longer stroke or a longer rod, or both, the piston's compression distance must be reduced. If stroke increases by 0.500 inch, there will be half of that, or 0.250 inch less available for the combined piston compression distance and rod length. Spend a little time and look through the catalog of your favorite piston maker; you will find either a piston that fits or one that you can special order at a custom price tag.

One way to measuring the piston compression distance is to insert a pin part way into the piston: now use a dial caliper to measure from the flat deck surface down to the pin. To this size, add half the pin diameter; for instance, with a 0.927 pin, add 0.4635. Compression distance gives you control over how high the piston is in the block and how much squish height—deck clearance—it will have. If an off-the-shelf piston is available with the right compression distance, you can equalize all squish heights with a little milling at the flat part of the piston deck. Decking the block can now fine tune the piston positioning.

Gas porting helps increase the expansion forces on the top ring to make it seal at high rpm. You can use this PMS fixture to hold and index the piston and to drill it. The center drill in the mill marks the locator holes so the drill will not wander.

This PMS vise clamps to your mill and is located by keys that make it parallel to the bed. Swivel block can be clamped at any convenient valve pocket angle.

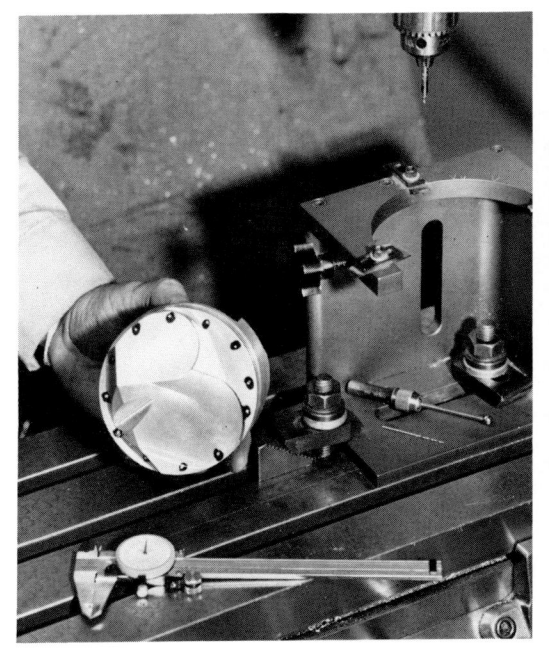

PISTONS - WRIST PINS • 21

RING TECH

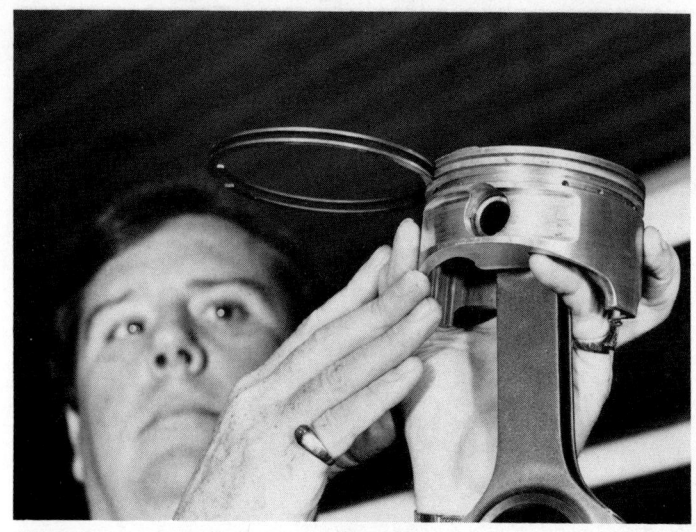

There is magic in those rings. Rings should be a smooth fit in the groove, and be able to roll all around.

This SS 50 Speed-Pro oil ring set is well down into the pin area on the Earl Gaerte engine. The pins are retained by buttons which also form the lower part of the ring groove, excellent for a big inch engine.

Rings make power. They also keep pistons alive by transferring heat to the cylinder. Among other duties, they prevent blow-by, keep out oil, and can travel at 5000 feet per minute. To sort out the best ring combination and gain installation fine points we touched base with Speed-Pro engineer Cal DeBruin and with engine builder John Lingenfelter, who runs an immaculate shop in Decatur, IN, equipped to build fast Chevy power. We also received considerable help from Earl Gaerte of Rochester, IN. Grab your helmet, we are going for a rapid ring ride.

Rings differ from application to application—the rings that fit your race car tow truck are not necessarily the same as the ones you want for pro racing. A normal street performance engine with a 6000 rpm limit will have 5/64 top rings to match existing pistons. This puts extra metal in the ring so it stands engine pounding for several years without being looked at. That size is also popular for turbo use where rpm is moderate but cylinder pressures increase with boost. Race rings range from 1/16 (0.062) to 0.043 and 0.039 inch.

TOP RINGS

The slightly rounded barrel face of the top ring accommodates piston rock. As miles pile up, the narrow contact line along the center of the barrel faced ring widens. In a Speed-Pro set, the top ring is usually made out of ductile iron—a tough chrome vanadium alloy which resists pounding and high pressures. Ductile iron is tough and can take some unbelievable bending.

Ring sets are available both in standard sizes, ready to install or oversized to allow for fitting to the bore. If you have just freshened up an engine and rehoned the bores, each 0.001 bore size increase would open the gap 0.003 inch and ring fitting compensates for this.

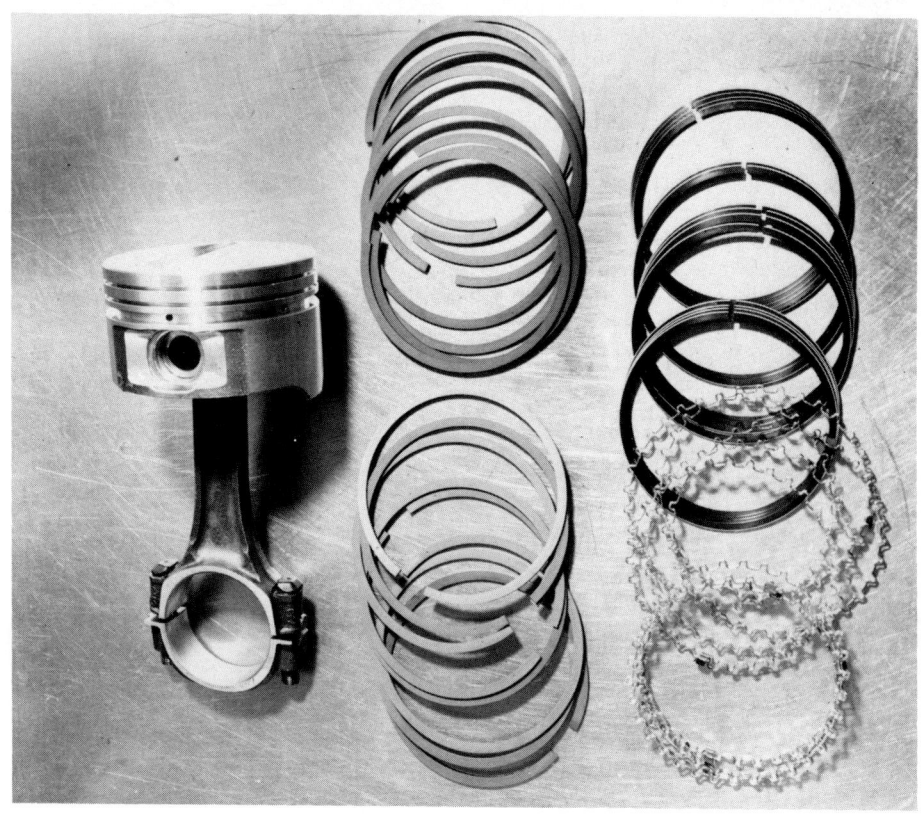

The ring should be set down from deck surface and squared up. Depth mike is an easy way to achieve this.

Chances are this ring will have a plasma applied moly face coating for good wear and durability.

A small bevel on the upper inside corner is designed to cause uneven tension so the ring will dish up and cock in the top groove. This seals the top of the ring groove to the top of the ring during the intake stroke. When gas pressure is applied during compression and combustion it flattens the ring so it seals against the bottom of the groove.

SECOND RING

The second ring does different work than the top one and has other design features. It is there to complement the top compression ring and to assist the oil ring. It is normally made of cast iron, has less tension than the top ring, and also less drag. To assist oil control it has a taper face—larger at the bottom than at the top. The taper face can ride up over the oil film on up strokes without driving oil into the cylinder, but then drives the oil downward—back to the crankcase—which helps the oil ring do its work.

The second ring has a narrow contact which assists with break-in. This contact widens with running, but by then the ring is seated. Both compression rings have dot markings on the top. If you find a second ring installed so the shiny wear line is at the top and the dot is on the bottom, you have just solved the oil pumping mystery—it is in backwards.

A second ring has an RBT designation which stands for Reverse Bevel Taper face which works exactly opposite the top ring: in fact, it seals against the lower outer ring groove corner for oil control and has good downward scraping action. Most second rings are run with uncoated faces, but for situations where more than normal wear occurs, plasma moly rings are available in several diameters.

OIL RINGS

The oil ring is usually a three piece

Check the ring gap with a feeler gauge. Top ring runs hotter and needs more clearance. You should run 0.004—0.005 gap per inch of bore in the top ring. With a four inch bore, that is 0.016 to 0.020.

SS 50 in a 3/16 width. Unless you are totally out of space and need a 1/8 ring use the wider more stable combination. There are two barrel faced hard chromed rails. The top one scrapes oil into the oil ring groove through the return holes and may also assist feeding oil to the pins when the piston is drilled for it. The lower oil rail returns oil through the clearance between the piston and the cylinder. We have seen Cosworth pistons with oil drillings half on the lower surface of the oil ring groove, something worth experimenting with.

Carefully controlled slots in Speed-Pro's stainless steel expander transform it into a spring. The expander pushes out against the rails and also applies them against the upper and lower sides of the ring grooves. Low tension expanders reduce drag in a race engine, while higher tension prevents oil consumption. Usually you want some compromise.

Where a mild street ring measures 5/64, a performance ring will be 1/16 and the reduction in width lowers tension, drag, and weight. The ring becomes more conformable and doesn't slam as hard against the ring lands. As you go up in rpm past the 8,000 mark, check into the thinner 0.043 top ring. By comparison a 1/16 ring is 0.062 wide. Tension (and ring drag) goes up as the cube of the wall

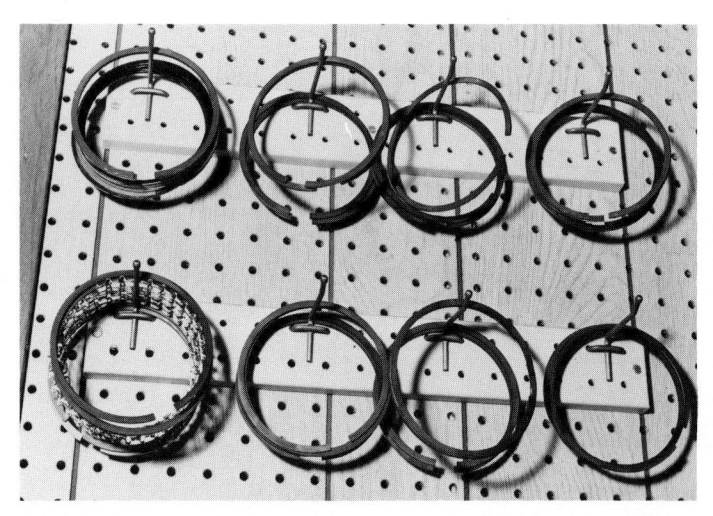

LAB Machine stacks their ring sets on numbered pegs, ready for installation. Anything you do to simplify and fool-proof the assembly avoids costly errors.

The top of the ring has a bevel which forces it to curl when it is compressed. This is a top ring, as you can see by the pip mark and the top bevel location.

thickness (how deep the wall is). A standard D wall thickness is arrived at by dividing the bore by 22. For instance, a 4.000 bore divided by 22 calls for a D wall of 0.182 inch. You can order a back cut second ring with less wall thickness to save drag.

GAS PRESSURE

Gas pressure has a great deal to do with applying the ring against the piston. To reduce drag, you need the least gas pressure during intake and exhaust; however the ring seal calls for more pressure during compression and still more during power. Making full use of the gas pressure is a story unto itself. One approach is Dykes rings which have a thin L-shaped cross section—very responsive to gas pressure—less wall stiffness, less weight, and more sealing. They have durability limits. Gas ports are another way to increase gas pressure—more for drag racing than for street.

You can drill gas ports from the top of the piston to the back of the ring. When this creates problems with class rules you can drill half on and half off the top ring land, radially in from the side. It meets the visual requirements and from everything we hear it works. Reduced stiffness of a 0.043 top ring plus gas ports are a popular combination.

If class rules call for running available pistons with a wide 5/64 groove you can convert with Speed-Pro's GI-60 to PBS-l3 spacers which narrow the groove width and allow you to use a pressure back ring.

RING DRAG

Ring sealing makes power. Reducing ring drag makes some power but if you go overboard you lose more than you gain. If you are inclined to reduce ring drag, help the engine by vacuum balancing the crankcase. Most of the oil is drawn in during the intake stroke when manifold vacuum pulls oil past the rings. If the crankcase pressure is low, less oil is going to migrate past the rings. You may be running down the strip with 1.5 inches of manifold vacuum and want the crankcase vacuum to be higher than that. Time spent working with the bevel of the pan vacuum tube in the collectors pays off, but if you don't measure it you don't know where you are at.

It takes time for pressure to act on the ring. If there is less volume in the groove behind the ring, the top and second ring react faster. The ring should be able to disappear into the groove and hide, but if it goes deeper than 0.02 inch you don't get full fast

This ABS ring gapper is state of the art with a diamond cutting wheel, a locator for squaring it up, and a dial indicator to see how much you have removed. This one is at Prototype Engineering.

24 • RINGS

Childs & Albert gapper fits into your vise and has two support pins to square up the ring. Touch up the ring corner edges with a honing stone to protect the cylinder wall during installation.

Cylinder bore geometry is critical, so we brought you a worst case example: there is no wear above the ring travel, very high wear at the top of the ring travel where pressure is maximum. Initial hone work at Gaerte Engines was done to show us that freshening up an engine may call for more honing than first meets the eye. If the air cleaner or its mounting has the slightest leak, the engine is sanding and you'll see more of this.

acting benefits. You can install ready made shims which go into the groove and fill the back of the ring space, speeding up the action.

Ring gaps have an effect on horsepower. If the gap is too large, it becomes a source of leakage. If the gap is too small, the ring can expand, close the gap and butt up solidly against the cylinder wall. The contact loading goes out of sight and the ring and cylinder wall suffer scoring, heat, and permanent damage. The tell tale signs during disassembly are shiny spots on ends of the ring. You will find in the Speed-Pro catalog an elaborate chart listing suggested ring gaps for many applications.

Rings don't live in a world of their own—they are part of a total picture. Oil is sprayed at them at the rate of 6—10 gallons a minute and naturally, the more oil you throw at them the more difficult it becomes to keep it out of the cylinders. If you control the amount of oil being thrown off by the bearings and reduce the windage in the crankcase as well as the oil on the cylinders you can begin to reduce ring tension and drag. Another point is honing—the surface finish of the cylinder has a large influence on the way rings bed in and seat.

What most people do not take into account is that the ring must seat in the piston groove and if that piston groove is of dubious quality as happens on some pistons, the rings can't seal. It is hard to measure groove flatness, but you should check the finish, new and after running.

Considering the amount of care put into manufacturing and lapping the rings themselves and the mirror like honing cross hatch finish, it is a little hard to understand that anyone would seat them dry although some builders do. Ring engineers tell us it's a bit too uncontrolled for their taste. You do not have to drown a ring in oil, but it does need lubrication right from the start.

If you do dyno work you will probably have occasion to compare rings to each other. Cal DeBruin tells us that on a block that has been freshly honed you let the first set of rings seat. From there on run your test and swap rings without rehoning. The rehoning would introduce more variables than it would solve.

The largest single item affecting rings is hidden—we are speaking of cylinder wall thickness, straightness, and even cooling. For instance, if you have a choice of running a 400 block with siamesed bores on the street or running a 400 crank with a 350 block as John Lingenfelter does in coming up with this 383, it is easy to see that geometry and wall thickness does more good than a few extra cubic inches.

RINGS • 25

BALANCING

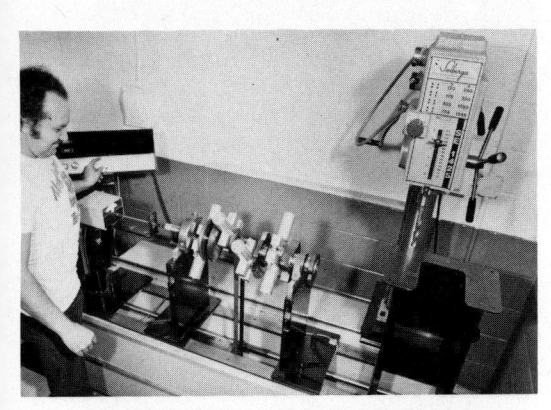

A government balances its books by borrowing more money. Real world engines don't tolerate that kind of balancing—they simple shake till they break. Apart from making the engine smooth and more pleasant, balancing is a must for durability. Let's take a V8 small or big block and see what some of the unbalance forces are and how to get rid of them.

If you tie a rock to a string and swing it, it will tend to pull your hand around with it, due to centrifugal force. The longer the arm about which the unbalance occurs, the higher the force. The crank pins, bearings, and the bottom half of the rods act just like weights swinging on an arm, and balancing them is a set of massive crank counterweights.

Next, let's consider reciprocating unbalance. We get the piston started from bottom to top and it wants to keep right on going because it has inertia and a sizable amount of stored up energy. The counterweights are designed to have some extra mass so that they can cope with this reciprocating force. Then, the piston reverses direction, again stores up energy but the crank must slow it down and stop it at bottom dead center and accelerate it the other way. When the piston is down, the crank weight faces up, so some of the forces cancel.

Things get a little more complicated because the rods, in addition to moving up and down, also swing from side to side. In fact they get complicated enough that you simply can't balance a single cylinder engine for all the forces involved.

There are reasons people don't go over 2.5 liters in a four cylinder without adding balance shafts. For instance, the average center of gravity of the pistons moves up and down twice every crank turn, creating enough shake to make parts fall off. In 90 degree V6s, the

A crank is balanced in two planes, one end at a time. When in balance at each plane, the entire crank is also in balance.

With a stroker crank, you may need many slugs of Mallory metal. They are pressed in from the side of the counterweight so as not to sling out from the crank.

Bob weights are made up to simulate the weights of the rotating and reciprocating parts of the assembly. They are centered on the journals and are horizontal when the pin is up.

Piston weights must be equalized together with the pin and rings. A little grinding at the pin bosses does it.

All weights are matched to the lightest, at the small and at the large ends of the rods. Rod is suspended by the big end, while the small end is weighed on an electronic scale.

Purchase and balance extra pistons to have spares on hand.

26 • BALANCING

A damper is balanced on the mandrel. The one exception is the Fluidampr, which comes ready balanced and cannot be spin balanced.

A moderate amount of weight can be added with welded in slugs or taken out by drilling. Grinding makes minor adjustments

front and rear of the engine does an independent hoola hoop. V8s have pretty good balance and surprisingly enough, so do in line sixes. A V8 with a 180 crank can put out more power than a V8 with a 90 degree crank, but is difficult to balance. Easiest way to balance an engine is to add external counterweights at the crank damper and at the flex plate or flywheel. This involves the smallest amount of weight and the least cost. It works on a street engine, but at high rpm, the engine falls apart because very high uncounterweighted internal forces flex the crank and kill the bearings. That is why you see race engines built with internal counterweighting.

The balancing job begins with equalizing the weight of all the pistons and pins to that of the lightest one, by using a precision scale. The rings and clips are assumed to be of equal weight or are weighed with the piston.

The next move is to equalize the weight of the rods and this is done by separately matching the weights at the big ends and then the weight of the small ends. This is important, because the big end is part of **rotating** weight around the crank pin, while the small end is part of the **reciprocating** weight and moves with the piston. The rod hangs from a pivot supported by a floating cable and is weighed at the opposite end. After all the light ends are "weighed", the big end weights are also equalized next. A little grinding at the rod balance pads removes the excess weight, matching up to the lightest one.

If you set a V8 crank into a balancing machine and spun it, it would literally jump out of the saddles. Point is that the crank has a considerable amount of counterweighting designed to compensate for the reciprocating action of pistons, pins, and the upper ends of the rods. Before the crank can be spun on the balancer, special bob weights are built and the correct weight clamped to each rod journal. The formula for the weight of a bob weight in a V8 or a V6, where two rods are run from each journal is simple and easy—it is the total of **twice** the rotating weight at the bottom of the rod, plus the reciprocating weight of piston, pin, clips, rings, and the upper end of the rod.

Crank balance is achieved by drilling out at the heavy spots on the counterweights, as indicated by the balance machine. It takes a combination of a good advanced machine and an experienced operator.

There are limits as to how much drilling you can do, and if the unbalance is large enough, particularly in a stroker, extra weight will be needed at the counterweight. To achieve this the crank is drilled and slugs of heavy Mallory metal are driven in. The alloy weighs more than the steel it replaces, which puts more weight in the same space. These slugs of Mallory metal should not be drilled in radially, if at all possible. Instead, the counter weights are cross drilled and the weights are pressed in. Now the centrifugal force is handled by the body of the counterweight rather than by a thin weld.

Flywheels are either balanced separately on a mandrel or are bolted to the freshly balanced crank and the assembly is then rebalanced only by working on the flywheel. This way if you replace a clutch or a flywheel, you only need to balance the replacement flywheel—the crank is already done. Pressure plates are handled in similar fashion by bolting

Flywheels, just like dampers, are spin balanced on an arbor. Install and check for run out with a dial indicator.

The center of the crank journal is lightened, reducing the amount of counterweighting needed. Journal is bored off-center to allow more metal around the oil galleries.

them to the balanced flywheel. Clutch bolts must be a tight fit in the pressure plate cover, or the pressure plate will not bolt up consistently. Torque converter flex plates are handled just like a flywheel. Crank dampers with a rubber insert are balanced on a mandrel, essentially as a single plane assembly.

For '86 and later small blocks, Chevrolet introduced a new one piece seal which rides directly on the end of the crank. This left no room for the external counterweighting, which has now moved to the flywheel or to the flex plate. That is why all late style cranks must be balanced with a flywheel or flexplate attached.

The one exception to all the things we have told you about balancing is the is the Fluidampr crank damper, which runs in balance on the engine, but cannot be balanced on a conventional balancing machine. Let us explain this riddle. On the engine, the crank torsionals create shear forces within the silicone causing the ring to self center. Since all the parts of the damper are factory balanced by Vibratech, the Fluidampr runs in balance. When you take the same unit and spin it up at a steady speed on a balancing machine there are no torsionals, so the ring rides to one side of the housing and shows unbalance.

BALANCING • 27

GASKET SCIENCE

Fel-Pro Performance head gasket is available with a steel ring for cast iron heads, or with a copper ring for aluminum heads. The copper ring has a larger bearing area and deforms easier than the steel. Figure a 0.038 gasket thickness and 9.7 cc in a big block and 8.7 cc in the small block.

Use a strong distributor gasket or a mag gasket that will not pound down and cause the timing setting to loosen and change. Also check distributor height and pan rail height before final assembly.

The driveway tells the tale—if it is dry, it is proof positive that you did the right things to your engine in terms of sealing it up. Getting a good seal with no leaks, oil drips, or air sucking into the wrong places is a partnership betwixt you and the folks who make the gaskets. Your end of it is to provide flat surfaces and the right studs or bolts, but theirs is to produce a good sealing gasket. Sounds easy, but it isn't, so we enlisted the help of some of the best racing people at Fel-Pro, including engineer Jerry Rosenquist and race manager, Roger Friedman, and we are going to help keep problems from happening in your engine.

VALVE COVER GASKETS

A valve cover gasket is constantly washed with oil and can see either vacuum or pressure. You can seal them with a cork gasket, but cork alone doesn't always hold shape. A better bet is Fel-Pro's Blue Stripe® material, a mix of cork and rubber. It has cork's resilience but is more stable and durable, the choice of most racers. Another option is Fel-CoPrene®, a type of synthetic rubber which is somewhat stiffer and works best with a rigid valve cover.

In any valve cover installation use load spreaders. With the cast aluminum covers you may have less room and may need to modify the load spreader or washer. The Vee shaped tips at the ends of the load spreader face down, so when the nut is tightened on the stud, the load spreader acts as though there were two bolts drawing down, narrowing the spans between pressure points. If you need a little extra height for valve cover clearance at the stud girdle or at the manifold, use Fel-Pro's Cork-Lam® which is a sandwich of two cork rubber gaskets bonded to a steel shim. The steel core stiffens the gasket, gives it a precise shape, keeps it from bulging out when tightened, and insures a superior seal.

To gain a flatter surface on a valve cover that has been used before, scrape off the old gasket material, then straighten the flanges where the bolts have deformed them. Next, apply a coat of Fel-Cobond quick drying cement to the top of the gasket and another to the valve cover, let sit for about a minute, and bond them. The other side of the gasket will rest against the clean surface of the cylinder head. This way you can remove the valve cover several times, reusing the same gasket. If possible, replace the bolts for the valve covers with studs which are locked and sealed in place with blue Pro-Lock I (where larger studs are used, retain them with the stronger red Pro-Lock II).

OIL PAN GASKETS

You have a choice of several materi-

Check the height of the front pan rail against the gasket seating base. Gaskets should be compressed by about 20 per cent of their free height.

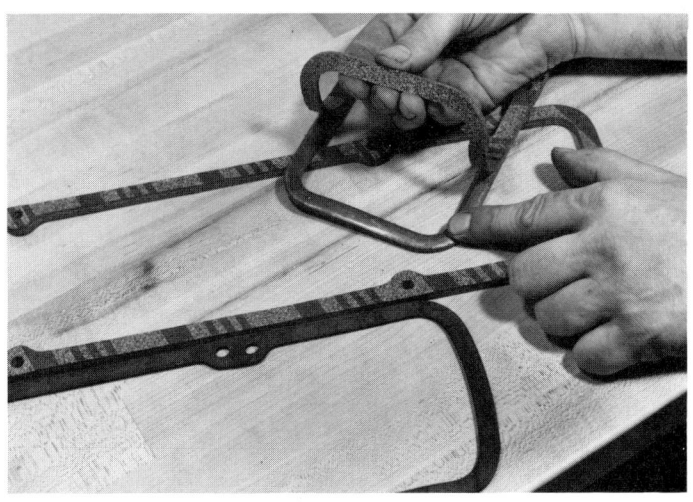

To raise the rocker cover and gain some extra clearance as well as improved sealing, use Cork-Lam which consists of two cork gaskets, with a steel shim laminated between them, to help the gasket resist side loads, due to vacuum or pressure.

als for oil pan rails: cork, cork-rubber, or Felcoid/Plus® which is a mix of tough fiber encased in a latex rubber coating and the outside is also given an anti-stick treatment. The latex outer conforms to metal surfaces but the center does not crush and remains stable.

Cement the pan rail gaskets and the front and rear rubber end seals to the block with a quick drying adhesive such as Fel-Cobond. The front and rear oil pan seals are cemented in place. Next add a touch of black RTV sealer at all four gasket/seal intersection. Early front seals from '57 to '74 are thinner than the late ones from '75 to '85. To

The oil pan rail needs to be straightened for each installation. Recheck the pan with and without the end rail gasket. A thicker front rail gasket is used from model year '75 and up.

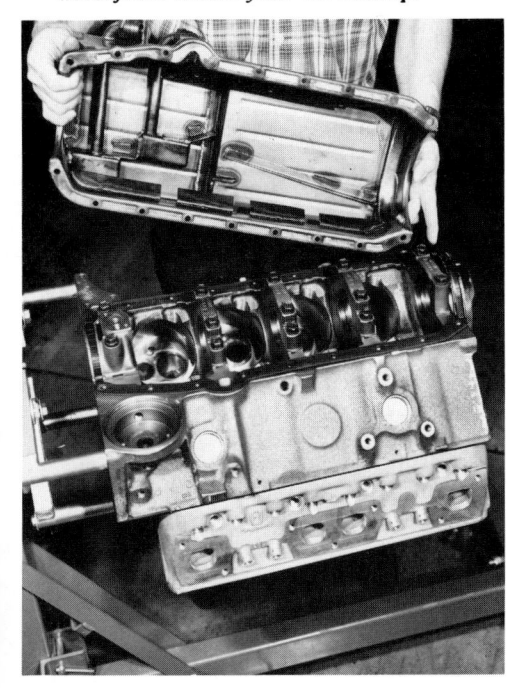

identify which pan you have, simply place a straight edge across the top front of the pan, then measure from the center of that straight edge down to the pan rail. On the early ones you will read 2 3/8" and on the late ones 2 1/4". Late style one piece pan gaskets, '86—'89, do not fit early blocks. When you snug down a set of pan rail bolts, don't overtighten and bend the pan rails. All gaskets eventually take a set, and now the bolts feel a little loose. Retorque them, instead of overtightening on the first pass.

Avoid silicone as a means of positioning gaskets, particularly rubber front and rear pan seals, because they tend to squirt out under pressure—the silicone lubes them instead of holding them in place.

WATER PUMP—TIMING CASE COVER—FUEL PUMP

With the engine running and hot there is a substantial amount of water pressure in the block. To avoid leaks, coat the gaskets with Brush Tack® sealer, put them in place, and also coat the pump side of the gasket. The same goes for the thermostat gasket. The Brush Tack contains filler materials, remains pliable, and does not squeeze out of place.

Where the engine is suspended from a front plate, instead of conventional motor mounts, and is retained by water pump bolts, it is a smart move to have the plates O-ringed on both sides to retain a better seal. There are also times when the belt loads on the water pump are high and you may want to use the Fel-Cobond instead of Brush Tack as a way of positioning the gasket more securely.

Fuel pumps involve two gaskets: one behind the adapter plate at the block and one in front. Since mechanical fuel pumps are seldom changed, use a Brush Tack Sealer on both sides of each gasket.

In a street installation that will not be taken apart, the timing case cover needs a coat of Brush Tack sealer on both sides of the gasket. The exception to the rule is a drag motor on which a considerable amount of track testing will be done. It then pays to cement the timing case cover gasket to the block with contact adhesive and oil up the front of the gasket, so it will not bond to the cover. You will need a small

Plain cork is not as good a sealing material as the blue stripe cork and rubber mix.

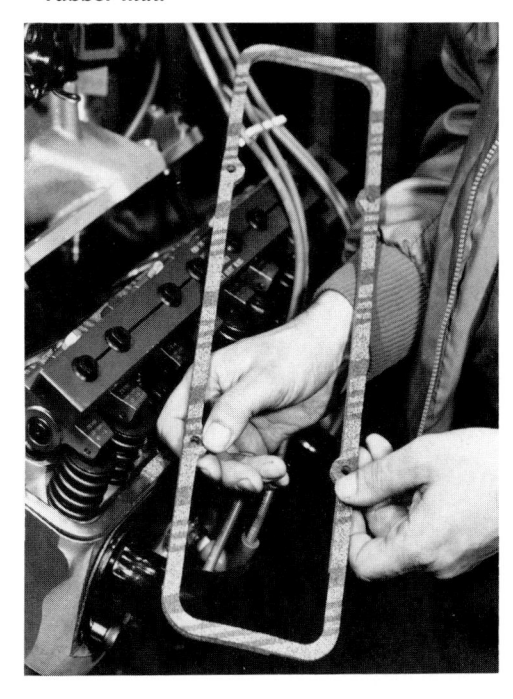

GASKET SCIENCE • 29

The world of chemicals never ends. Some Anti-Seize applied to exhaust bolt threads keeps them from freezing or breaking. Use metal retainers to keep them from loosening. Aluminum epoxy can fill in gouges or act as a buildup to improve flow.

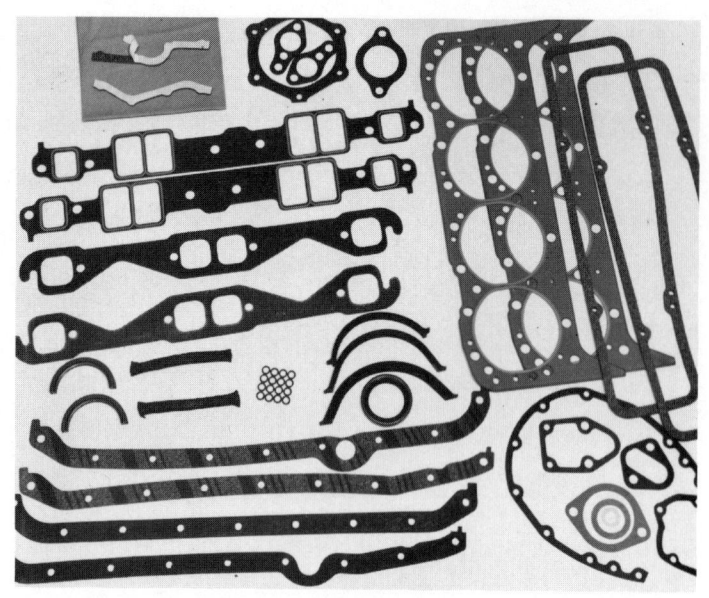

Each gasket is a study in materials and design features. 1004 and 1010 head gaskets are cut away at the ears to make room for rev plates. You can use either intake manifold rail gaskets or evenly applied silicone beads which squash down and form a saddle over the rails.

amount of silicone at the timing case cover base where it contacts the pan.

MAG AND DISTRIBUTOR GASKET

With a poor quality paper gasket under the distributor, race shake compacts the gasket and now the distributor loosens and turns, causing timing changes. Since a mag is heavier, the shake is greater and it loosens even sooner. Race engine builder Earl Gaerte mentioned this to Fel-Pro and they solved it with a rubber fiber gasket that holds its thickness and will not allow the mag to loosen. R.H.S. has a neat Mag-Lock which clamps in place around the mag and has an adjustable link. Now you have the best of both worlds.

INTAKE GASKETS

Because a manifold is a wedge, it tends to drive itself down between the heads. You tighten the bolts and the casting turns in at the top but pulls away at the floor of the ports. Now the manifold leaks oil and sucks oil from the valley. Fel-Pro has measured compressed thickness of used intake gaskets and found that can be 0.005 to 0.010 inch thinner at the top than at the bottom because of uneven loading.

There are also many machining tolerances. You should check the manifold against the cylinder head template or intake gasket. Do a little heliarcing on the manifold if the sealing surfaces are too narrow.

If the manifold has been shaved before and is sitting low so that the manifold port floor is below that of the head, you can often save the day by switching from the single thickness 0.060 intake gasket to an optional 0.125 double thick gasket. This raises the manifold and realigns the ports.

There are several optional intake port sizes such as a Fel-Pro 1205 for a modified Bowtie or an LT-1 head, or a 1206 which fits a short track or a Winston Cup head. There is a larger 1207 for drag racing, sprint cars, or long track Winston Cup cars. The 1209 is sized for big cube motors with number 12 Brodix ports used on drag and sprint motors.

Most of Fel-Pro's performance intake gaskets have thin lines of silicone rubber—Printoseal—beads around the intake ports and coolant passages. This provides extra compression in the sealing area plus better bonding and conformability. There is also a 1204 gasket for street use, designed to seal rigid factory cast iron intakes.

HEAD GASKETS

A head gasket operates under difficult conditions. As soon as you start the engine, each combustion chamber in turn sees a high enough pressure load, to make it want to lift away from the gasket. If those forces can unload the bolt clamping pressures, high temperature combustion gases leak by, which will "flame-cut" any gasket. Using high grade bolts or replacing bolts with studs and using hardened washers under bolt heads or nuts will prevent many problems.

To perform their function, head bolts or studs must stretch. You tighten the fastener and the threads are going in deeper, the head is not compressible, the bolt head is staying put, and so the bolt shank stretches and applies a load. After the gasket has been in operation, it compresses permanently, and takes some slight amount of set. Now the lower stretch action reduces the bolt loads. It is not that the bolts backed out, the set of the gasket loosened them. On a race engine the heads are re-torqued to make up for the set. The long studs on the Chevy head stretch more and retaining tension is not a problem, but short bolts stretch much less for a given torque and loosen more readily. That is the reason for undercut short studs—the thinner center section allows more stretch and you make up for the difference in diameter by using higher alloy material with greater tensile strength.

Today's modern Fel-Pro gasket has a steel center core to which are bonded rubber fiber sheets, forming a sandwich. Around the combustion chamber edges there is added reinforcement in the form of C-shaped steel armor. To this basic high grade gasket is added a blue teflon coating. It improves the seal between the gasket and the flanges. Some race head gaskets, just like intake gaskets receive a coat of Printoseal—raised beads of sealing material around the cooling ports. Those beads flatten into the gasket when the heads are torqued down, to help with sealing.

Before you even put in the gasket, bolt up the header and make certain that it is straight and can seal. These coated Fel-Pro gaskets will not stick to the header and are reinforced with an inside steel core.

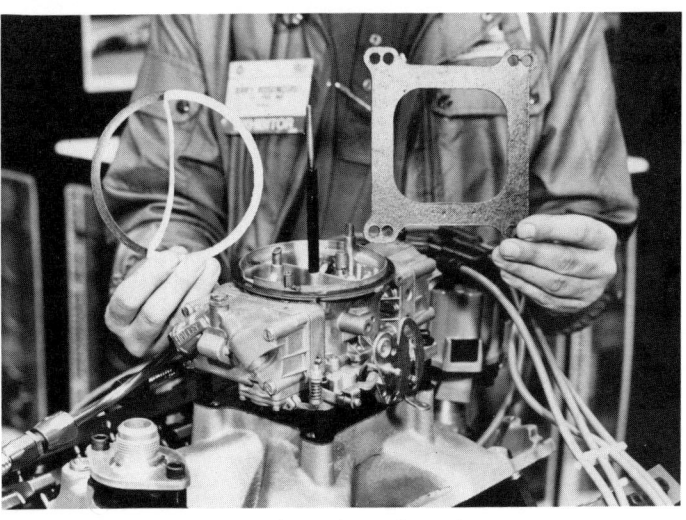

To prevent air cleaner leaks and sanding the motor, use a gasket at the top of the Holley air horn. This Fel-Pro 2104 has an adhesive base to permanently install it on the air horn and the steel core to protect the carburetor flange.

Keep in mind that the head and block are constantly expanding and contracting, so no matter how rigid we make them, the gasket still needs resilience to follow the shape changes.

In a conventional street type gasket, the armor forms the protective band around the combustion chamber edges to keep combustion chamber flames from burning into the gasket. In addition, the double thickness of the flange, around the gasket increases the local sealing load around each chamber.

For race engines Fel-Pro designed a series of gaskets with steel wire rings positioned on the inside of stainless steel armor. This drastically increases the clamping action. In a small block, the 1003 will fit up to an over-bored 350 engine or a stock 400. A 1004 accommodates a 400 block with a 4.125 bore.

When you use these gaskets with steel heads, they work great. Now you try them on an aluminum head and the high unit load can cause the wire insert to Brinell a ridge in the soft aluminum head. The metal moves as on a bad U-joint where rollers have dug ridges. Here again Fel-Pro came through with a neat answer. They designed a 1010 gasket for the small block and a 1027 for the big block which have a larger wire copper O-ring that has been pre-flattened. Now the load distribution agrees with the aluminum and you do not get the wire imprint.

Girls like a big hug and so does a gasket. To gain a good clamping action, a rigid span is needed between the head fasteners, so the less you cut from the deck in the head or the block, the better the gasket will like it.

The Surgeon General advises that standing on the edge of a cliff is even worse than smoking, because edges fall off suddenly. By the same token, positioning a gasket at the edge of a cylinder bore causes that edge metal to fail. You want a little safety rim, like having the gasket 0.015–0.020 per side away from the edge of the chamfer on the head or block. Now there is enough surrounding support metal and the gasket remains firmly clamped. Spend the extra time and scribe the location of the cylinder bores while the heads are bolted to the empty block and ask your porter to please not fall off the cliff. The inside opening of the gasket will increase around the bore when compressed, so do your marking with a used gasket.

As the engine receives larger bores in succeeding model years, very little material is left in the bridge areas between the bores. This causes gasket headaches because of increased temperatures and lack of coolant. Brodix designed their Spread Port aluminum heads to move the center exhaust ports away from each other and to make some room for cooling. Race engine builders also recommend, for some applications to plug the small steam holes next to the bridge area so a crack does not originate from that point.

The front cover seal should receive a light coat of Brush Tack before pressing it in. You have a choice of gray twin lip or black single lip rear main seals. The gray silicone seal is preferred.

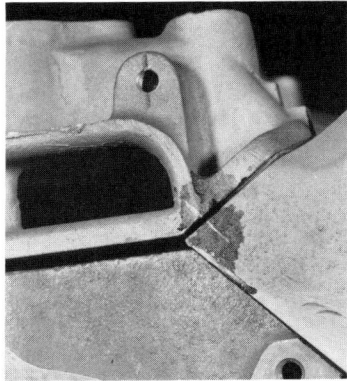

A wide range of gasket port sizes are available for both intakes and exhausts for different porting levels. Intake manifolds often need an extra weld bead top and bottom.

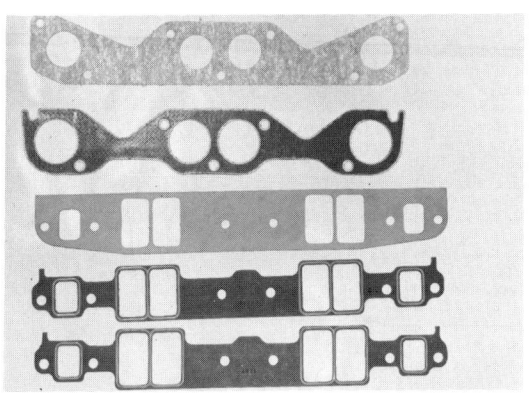

GASKET SCIENCE • 31

PREFITTING

Use a tap in moderation—you want to clean out the threads not auger them out. A new tap removes too much material from the block threads.

Four-bolt mains with angled bolts provide a better tie in for the block. The standard outer bolt holes remain unused with these angled bolt Diamond Racing Products caps.

If you had a staff of NASA scientists working for you, you'd incur a few postponements in your engine launch. A racer or an engine builder assembling a top notch engine may spend a winter of premeasuring, fitting, and machining. A street engine shop may throw their motors together overnight. Who is right? Actually, all of them—if you want to innovate, build something far out—it takes time and money.

Here, we're going to hold an eye-opening prefitting session, which must be better than hoping that the parts out of the box will fit together. Yet, we're going to keep cost in mind and give you a balanced picture.

BLOCK DEBURRING

Block deburring is a job that can last an hour or a day, depending on how thorough you want to be. It can mean removing all sharp casting edges to avoid cut up hands and smoothing down inside casting surfaces to remove surface dirt and improve oil drainage. It can save some block weight, depending on the application. You need a flame shaped carbide burr, same as for porting heads and a good supply of air. While you are at it, put a funnel shape at the oil drain holes in the head, match the oil drainage between the head and the block, and improve the drain back from the lifter galley.

BOTTOMING TAPS

You can take advantage of the parts not being any too clean to run a bottoming tap through all of the bolt holes in the heads, at the intake and exhaust manifolds, and also through all of the head bolt holes in the block.

A bottoming tap differs from a conventional tap in that it does not have a tapered lead in section. Its nose is cut off square, which allows the bottoming tap to reach a little deeper, completing partial threads so that bolts will not jam. Too much vigorous tapping removes too much metal, and weakens threads, so use well worn taps for clean up.

LIFTER VALLEY OILING

All of the oil delivered to the heads comes cascading down to the lifter valley. It will save horsepower if you can avoid having this oil drain down directly over the crank, adding to the windage. To achieve this, many engine builders tap the oil drain holes over the cam and plug them. Some engine builders use stand pipes instead of those plugs. The idea is to let the crankcase breathe through the stand pipes while the oil drains front and rear. Breathing and oil

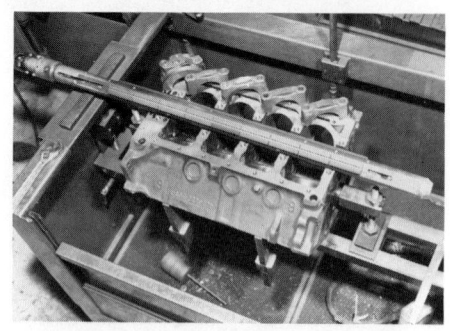

A two bolt block had ample strength and is easily converted to four bolt use. When new caps are installed, they must be line bored and then line honed.

drainage can interfere with each other if you have two-way traffic.

Debris and metal particles from valve train wear comes down from the heads and you don't want that stuff getting into the crankcase. The best defense is some stainless steel screening epoxied at the outlets of the tappet valley, available from Moroso or similar sources. Install those screens as a last step, when the block is ready for assembly.

TWO- AND FOUR-BOLT MAINS

You have a choice between two or four-bolt Chevy blocks, and for any serious horsepower over the 300 mark, you are better off using four-bolt mains. They are readily available at wrecking yards at an extra price and worth the money. If you are building an under 5000 rpm engine with hydraulic lifters, capable of 250-350 horsepower, two-bolt mains will easily stand up. It is the

With a dry sump or a wet sump with an external pump, the stock Chevy rear main can be replaced with a steel one such as this Diamond cap.

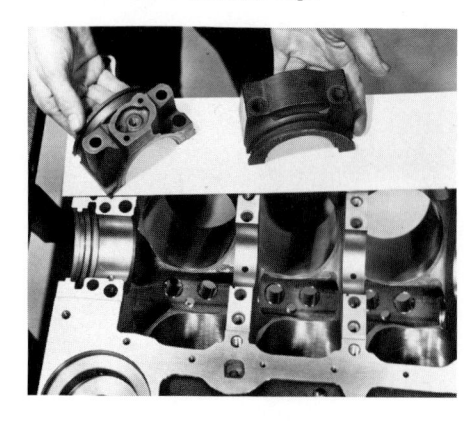

32 • PREFITTING

rpm, more so than horsepower, which forms the dividing line.

With a higher number gear in a bracket car, or a hobby stocker with a solid lifter cam, you can either use a stock Chevy block with four-bolt mains, a Bowtie with four bolts, or you can convert a two-bolt block to accept four-bolt steel after market caps. Most four-bolt caps from Batten, Summers, Milodon, and many others have splayed (angled) outer bolts which offer a good tie in to the outside of the block.

If you build a block with two-bolt mains, replace the bolts with studs. They engage deeper and make full use of the threads. When you convert those studs, finish the bottom threads with a bottoming tap and use hardened washers at the nuts. Normally, just the three center main bearing caps are replaced. Some four-bolt caps use studs at the center and angled bolts or hex head studs at the ends.

The rear main bearing cap carries the oil pump. You can replace it with an equivalent steel cap. If you go to a dry sump oiling system or combine an external pump with wet sump, you can either just plug the passages in a stock cap or go to a steel one. The number one cap can be reinforced with a steel block or be replaced with a steel cap.

The caps are located from side to side by steps machined in the block. New caps require precision machining at the parting face and so does the block. There needs to be a light side to side interference fit at the sides. After a cap is installed, some builders add small stake marks with a center punch to press block metal against them. New main bearing caps must be line bored, after which the block is ready for line honing.

To do any line honing on a 400 inch block involves an extra step. It has larger main bearings than a 350, but the same size seal. To make room for the line hone to go through, the seal support area must first be bored out. Now the seal saddle is too big and you make up for it by using an aluminum shim between the seal and its saddle. Some use tie straps as a shim.

In any block, when the rear main bearing is remachined at the parting face, the seal saddle is squeezed together by a small amount. To correct the inside diameter of the seal Bob Patzold from Prototype Engineering in Hainesville, IL has made up a cutting tool that mounts on an arbor piloted by the three last main bearing saddles.

In this Prototype motor, all of the cam drain holes receive stand pipes which allow venting and extend well above the oil level. Stainless steel wire mesh screens provide the oil return but prevent valve train trash from reaching the dry sump pumps. All of the lifter bores have been rebored and bushed.

This allows the seal bore to be corrected with considerable accuracy. The same arbor also carries a tool to correct the thrust face of the rear main. It's the kind of tooling that makes all the difference between a major engine builder and the round-the-corner machine shop.

The rear flange of the crank makes it impossible to see how well the seal fits. To avoid leakers, master engine builder Earl Gaerte from Rochester, IN built an elaborate rear seal inspection tool, using the back half of a crank and machining off the flywheel flange. The tool is installed together with the rear seal before the crank is laid in; now, you can see at a glance how well the seal will contact the crank. This eliminates leakers before installing the engine.

LINE HONING

After a new block has run, it will move around a bit and stabilize. Hot and cold cycles and the operating loads have seasoned it. However, moving around contributes to the main bearings going out of alignment. With a 0.002 or 0.0025 bearing clearance, you want the bearing bores straight and

Spend the time and match the oil pump outlet and the main bearing so that there will not be any oil flow restriction.

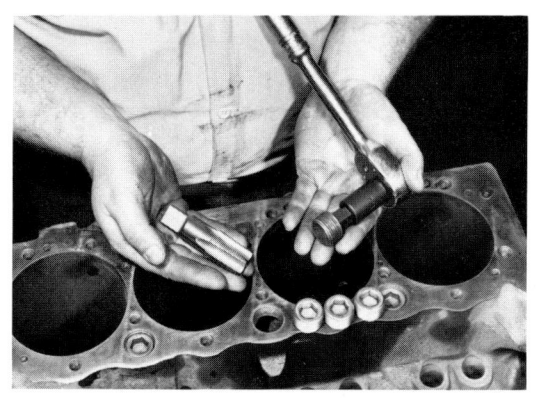

To strengthen the block, make room for some cast iron plugs at the three core holes at the top deck. Each plug needs a small coolant passage drilled in the center—it also serves as a steam vent.

the best way to cure the problem or to induce new ones, is line honing.

The Sunnen line hone has a long massive arbor with support shoes and stones. It is set into the main bearing saddle, the caps bolted on and torqued down. Since the arbor both turns and moves fore and aft, you produce round main bearing holes in perfect alignment. Keep in mind that the stone removes metal from the bore, but the metal is also breaking down some of the stone, causing a mutual truing action. Cutting oil floods the areas and washes away the grit so you have clean fresh cutting surfaces on the stone. You gain a cross hatch pattern good at holding the bearings and straight bearing bores in line with each other.

Now let's add a little more detail. If you simply put a stone in a block, as we described, the hone would work, but your would need oversized bearings. To avoid this, the main bearing caps are first cut down by a slight amount. Since most of the out of roundness takes place in a vertical plane, the stones now have some metal to bite into, and can correct the bores

PREFITTING • 33

Here the block is centered on a bar by the main bearings; top table of this Kwick Way boring machine is parallel to to the bar, so the bores will be perpendicular to the crank center line. You're visiting Prototype Engineering.

Before the block is honed, it is fitted with a deck plate. Use the same type of fasteners as for the engine assembly, to the same torque specs. You are at Jeg's with Keith Burton.

34 • PREFITTING

back to standard size. To put it simply, you squash the circle and then make it round again. The caps need just a little cleanup and we are dealing with only a few thousandths.

When you remove material from the cap and "squash the circle", the hone will bite equally into the top and bottom halves of the bore. This moves the crank center line up into the block by a few thousandths and now the timing chain is not quite as tight as it would normally be. The timing chain folks at Cloyes supply slightly oversized sprockets that take care of this problem.

B-H-J makes a fixture which measures the center to center distance at the block between the cam and the crank and gives an exact size for the chain and sprocket combination as well.

CAM GEAR ALIGNMENT

When you were a kid riding a bicycle, one of the early facts of life was that the chain had to be straight, or it would wear and jump. Now that you have graduated to race cars, the facts of life haven't changed; you still want the crank and cam sprockets to line up so the chain won't wear and die.

Our first step is to inspect the cam thrust surface. The helical cam gear which drives the distributor and oil pump applies a thrust which sends the cam and gear toward the thrust face at the front of the block. If the surface is worn, which is often the case, refinish, or do something about it. The minimum cure is a couple of steel shims which cut down on the wear action. Another good cure is a Torrington thrust bearing that has two steel shims and a roller thrust washer. It runs forever in a flood of oil.

At this moment, you have just become a smart engine assembler, installed the cam, gear, timing case cover, put a dial indicator on the tail of the cam, and said "I have 0.003 inch end play and I am going racing." Then on final assembly you use a cover gasket from another kit and the clearance changes. "My helper did me in. And I was leading the race on the 49th lap when it happened...."

To check if the gears are in alignment, install the cam and crank gears, place a straight edge against them, and it should be flush at both gears. if it isn't, you either need more shims behind the cam gear, or our need to chuck the cam gear into a lathe, indicate it, and cut just enough to bring the two gears in line. Iskenderian makes a great front cam bearing that sticks out far enough from the block to pilot the torrington shims and thrust washer. Gene Ohle of Evans Machine in El Monte, CA, built a cutting tool with an arbor that pilots in the cam bearing bores and cuts the thrust face on small and big blocks.

CRANK END THRUST

The rear main handles the crank end thrust (see the Crank Section). Every racer with a strong left foot, and a heavy clutch, makes it a practice to feel the play at the crank damper between races. If there is a sudden increase in play, the end thrust has increased and you need to do something about it quickly. Engine builders who supply additional oil for crank end thrust have fewer of those problems.

Replace any press-in plugs with screw-in plugs that will not pop out or leak like cup plugs, when you least expect it.

BLOCK DECKS

We come to visit Jeg's in Columbus, OH and Keith Burton is busy threading a big tapered tap into the top holes of cylinder block deck. Next, Keith threads in a set of tapered fine threaded cast iron plugs. When driven into the core holes at the top of the block, they strengthen the deck and keep it from moving around.

Strong block and head decks are major contributors to staying out of gasket leaks. The less you cut from them, the better for gasket life, unless you happen to be in search of compression and haven't found pistons of the right dome height or compression height.

After installing the deck plugs, drill small passages in those plugs to vent steam pockets and to allow coolant flow. Extra steam holes affect coolant distribution, so consult with your machine shop before randomly drilling holes. Also check the gaskets you just

If a block has been freshly cut, all bolt and coolant holes must be chamfered.

The block must be leveled so you get a zero indicator reading from one side of the block to the other. Each cutter is set at the same height.

pulled out to see if they need extra cooling in specific areas.

A small block Chevy has exhaust ports adjacent to each other near the center of the head and there is a good reason for people like Brodix making Spread Port heads where the center exhaust ports are wider apart. Brodix also has drilled holes and tapped connections which pull extra coolant from the center of the head and bring it to the radiator side of the thermostat housing.

The cylinder head locating dowels must be removed for block decking. There are neat tools for this, but in a pinch a claw hammer driven against the dowel grabs and holds. Later, you simply install a new dowel. Think everything is hi-tech? Clever engine builders will on occasion move dowels to bring the intake valve further away from one side of the bore or will overbore in one direction or another to improve engine breathing.

PISTON TO DECK CLEARANCE

In figuring piston to deck clearance, add the thickness of the head gasket. Say you are running a blue Fel-Pro gasket with a 0.038 inch compressed thickness. If your piston is 0.010 inch down in the hole, the total deck clearance is 0.048 inch. With an embossed steel gasket, standard street fare not really designed for performance, you would have less than 0.030 deck clearance with the same piston location.

The best way to measure piston to block deck height is with a dial indicator. Zero the indicator against the block and swing it toward the piston. Keep in mind that the piston to deck clearance varies a bit as the piston rocks in the bore. To check the minimum clearance you have to rock the piston up toward the indicator by pushing on it from the opposite side of the wrist pin.

How accurate does the deck clearance need to be? In a street 9:1 engine a few thousandths will make no difference whatsoever. You make them accurate, but you can never pick up the difference in performance. In a 13:1 or 14:1 small cubic inch drag motor every thousandths counts and you can never be accurate enough.

You have indeed squared up the block, but there are several other little variations to cause problems. For instance, not all crank throws will have the same stroke. Sometimes deliberate errors are introduced by grinding the crank a little off center to keep from going to the next undersize. The rod lengths may not be exactly the same—also, the piston height from the wrist pin up to its flat topped surface can vary. Now you start doing a mix and match job to equalize the piston to deck height.

A pair of calipers give you an approximate rod length. A depth mike or a caliper from the top of the piston to the wrist pin, gives a compression

A flood of oil cools the cylinder bore and washes away grit. The hone constantly overstrokes in both directions (extends beyond the top and bottom) to maintain straightness, and a mutual truing action.

Restrictor type screw in plugs can be used to limit lifter oil.

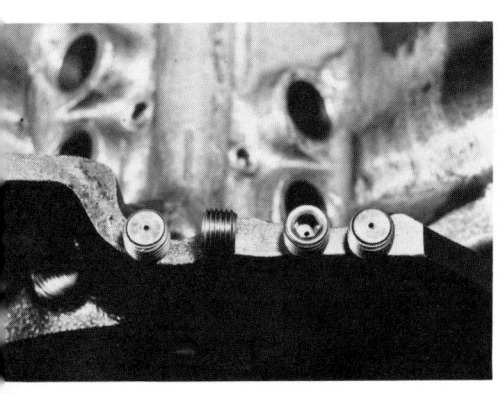

In the Storm Vulcan, the block rests on an alignment bar. The cutter will machine the top deck on each bank so it is parallel with the crank center line.

PREFITTING • 35

After honing you need hot water and detergent to clean out the grit and the block must be oiled up before the bores rust up.

Honing begins with coarser grit stones to reach size and finer stones follow, to achieve the desired finish. Since material is removed from both the stones and the cylinder bore, a mutual truing action constantly takes place.

height (add half the pin diameter). An engine builder like Diamond Racing will use a marble reference table and a height gauge for the same purpose. When building an engine with floating pins, you can change piston and rod combinations easily, then renumber the parts. On an engine with pressed pins, belt sand an old pin so it just fits through the rod, to allow fast tests.

CYLINDER BORING

Most people overbore cylinders too far and wonder why they lose power. Cylinder block castings are thin. When you deal with cylinder walls that are 0.125 to 0.090 thick there isn't much metal there in the first place. When bored paper thin, walls bulge under pressure and increase blow by; the less you bore, the stronger the engine.

Before you do bore, consult your Speed-Pro or equivalent catalog and find out what pistons you can get. Are you leaving a rebore or a rehone margin to save an expensive block? Rings are available prefitted, or oversized so that you can fit them after rehoning the block. Some special ring sizes are listed, but manufacturing runs occur at infrequent intervals, and getting them is something else. The same goes for odd piston sizes; check actual availability. Better yet, have them on hand before machining begins.

The cylinder bores should be square with the crank center line. Smaller boring bars set directly on the block and rely on the deck surface for accuracy. More sophisticated boring bars float on an air table parallel to the center line of the crank—less margin for error.

We come visit with LAB Machine in Lindenhurst, NY and Lee just happens to be setting up a block on his Storm Vulcan machine. A long square bar with rounded corners is dead level and the block rests on it along the main bearing saddles. The cutting head and its massive support arm swing parallel to it. The block is set with one bank up, leveled, and locked down. A Storm Vulcan has a large rotating head with cutters all set at the same depth. The block is indicated from side to side and along its length. The machine, as Lee points out, is as totally accurate as the setup. You want your decks parallel to the crank center line and squared up because this affects the piston to deck clearances.

HONING

Once a block has been bored, the cylinders must be honed to remove the fractured metal surface left by the boring bar tool. This is essential to good ring seating. Generally, 0.002—0.003 inch of material is left in place for honing.

Now a hone is capable of making precision bores in the hands of a skilled operator. However, just because the shop has a hone doesn't mean they know how or want to spend time to make round holes. Here are some of the things the operator of the hone is faced with and if you understand them, there will be better communication.

A hone removes metal by abrasion. It exerts pressure on the cylinder wall and scrubs off metal particles. Honing stones just don't happen. Sunnen makes them to amazingly accurate specifications from special abrasives that are powdered, graded by grit size, and bonded with special agents. Honing stones are purposely designed to continuously break down and expose fresh cutting surfaces. A stone which doesn't break down fast enough loads up with metal and stops cutting. Also, a honing stone must break down so that the cylinder wall can keep it straight. Hones must overstroke—extend beyond the cylinder bore above and below it—to stay straight, but they cannot extend so far as to hit the main bearing webs.

Since a hone exerts pressure against the cylinder wall, the unsupported middle of the wall is going to flex more than the ends. If we did nothing about it, the cylinder would be bell-mouthed at both ends because the hone would remove less material from the center than from the ends. If you hone a cylinder perfectly round and come back a

A simple driver helps install freeze-out plugs. You can add retainer screws to help them hold pressure.

36 • PREFITTING

few hours later, it is no longer perfectly round. The metal has stabilized, moved, and you can start honing a little more.

There is an entire science of honing stones and grits and finishes. Every machinist has his very own combination and some of them work much better than others. Generally, a coarse stone brings the cylinder to size, with finer stones applying the finish. Finishing costs money because it takes time—you can't remove the metal as fast with a fine stone.

Use high pressure cleaner to dislodge sediments from the water jacket and the galleries.

Gene Ohle explains that you can spend one hour or eight hours honing a block. At $40 an hour of shop time, your value judgment changes between a street machine and a Winston Cup car.

Every high school kid knows that you need a deck or torque plate bolted to the top of the block before you can create a round cylinder bore. That torque plate is needed so that you can apply the bolt stresses to the block and distort it just as though a head was bolted to it. Most machine shops buy a torque plate and use it forever after. Top engine builders, like Gaerte, check those torque plates every Monday morning to make sure they stayed flat, refinish them if need be and often get new ones. Engine builders, like R.H.S., frequently replace the fasteners that they use for the torque plates because threads wear from disassembly. They also make it a point to use the same type of fasteners that will go into the engine for final assembly.

END GAPS

If you straightened out a ring on your average small block it would be a foot long. It also sees a lot of heat. Now we need to accommodate the expansion factor and that calls for an end gap. You do not want this gap too big so gases leak past it, but you also do not want it too small. When there is not enough clearance the ring expands, and goes solid. Now it can apply a tremendous scrubbing pressure to the cylinder, enough to wipe out the bore and the ring finish. If you see shiny surfaces on the ends of a ring, there was not enough clearance, they were rubbing, and you were living dangerously. In a sharp drag engine, high ring gaps can cause blow-by and a slight power loss. This is nothing compared to the power loss you get from wiping out the rings from not having enough gap. The hottest top rings need 0.004 to 0.005 gap per inch of bore. Usually the top ring operates at 0.018 and the second ring at 0.016. Every thousandth of bore wear puts three thousandths increase on the end gap. To gap the ring bring it an inch down below the top of the bore, square it up with the piston or with a depth mike, use a feeler gauge, and check the gap. If it is not right use an end gap tool. It is very easy when you are doing your first job to form a gap that is not squared up. Think about it, center the ring and your gaps will come out square. The ends of the gap should be lightly stoned off to prevent rough edges and scratches.

BLOCK CLEANING

A block is not ready for cleaning until the cam bearings have been removed. Concealed behind cam bearings are oil feed grooves in the block. Equally important is pulling from the block all the soft oil gallery cups, drilling tapping, and replacing them with screw in plugs. Those plugs are installed later, as part of final assembly. However, part of prefitting is making certain that the plugs are not threaded in so far they will close off the lifter feed galleries. You can clean blocks by steam, vapor degreasing, water jets, with solvents, but as far as we are concerned, any system involving bead blasting or burning off deposits with heat, exposes your engine to total disaster.

Honing leaves grit embedded in the pores of the metal. The only way to get it out is with detergent and hot water. Solvents will not clean it out. Once the bores have been washed, oil them immediately to prevent rust. Put the block in a plastic bag to protect it against dirt, when it is not being worked on. Vent the bag so condensation does not form and rust the block.

Crank oil galleries can be cleaned with rifle brushes. A light touch with a ball hone removes varnish and burrs from lifter bores.

The hone head carries two stones and two shoes which are preset with shims, for correct cylinder geometry.

PREFITTING • 37

VALVE TRAIN GEOMETRY

When you install rocker arms, make sure they clear the valve spring retainers. To achieve this you may need to add pushrod length, a lash cap, or both.

This cutaway model at Competition Cams shows the pushrod socket and the oil spray gallery. A pocket above the pushrod insures that oil is delivered full time while the rocker swings through its arc. When you prelube an engine, turn the crank and make sure oil is coming through each of the pushrod and rocker assemblies.

You have a choice of 1.5 and 1.6 Comp Cam roller tip rockers—the engine will tell you on the dyno or at the track which one it likes best.

To check valve lift, make sure the indicator stem is in line with the direction of valve motion. This rocker arm is too far back on the valve stem, riding on edge. A longer pushrod is needed.

In the days of the first Corvette, Chevrolet had a six cylinder in-line engine with a long rocker shaft that ran just fine. Then came progress, a V8 small block with bathtub rockers and pushrods guided by the head. Then came screw in studs, guide plates, stud girdles, and finally a high buck set of Jesel rocker shafts and arms to return to square one. You've come a long way baby.

In putting this section together we received considerable information from the tech guys at Competition Cams on all the real world things it takes to make a valve train live.

We also spent some fascinating time with Jim Covey from Chevrolet Raceshop (Special Products). Jim set up a highly modified Optron on a cutaway block on one of the dynos at Katech Engineering in Mount Clemens, MI to study valve train motions under dynamic conditions. It was a straight case of math and advanced science making an engine go faster.

When you run stock valve springs, small valves, and a zilch cam it becomes very difficult to screw up the

38 • **VALVE TRAIN GEOMETRY**

On older Chevy heads with oval shaped pushrod slots and press in studs, R.H.S. uses a guide and a drill to open up the holes so they do not interfere with studs and bolt on guide plates.

You can easily see the angles formed between the valve stems, studs, and pushrods. Increasing the rocker ratio makes pushrods work harder.

valve train. Now you add larger valves, increase duration and lift, install pop up pistons to gain compression, and all of a sudden piston to valve clearance and valve train geometry become items you want to double check about six times.

When a valve hits a piston, the bending and breakage that ensue ranges from minor to catastrophic and you are distinctly out of the race. It is also common to be out of an engine, because broken piston pieces travel through the manifold and clean house on adjacent pistons.

The valve train throws hints of its displeasure by bending pushrods, breaking springs, showing marks where the valves kiss the piston and speaks with many other little signs. The best time to keep all this from happening is during prefitting. Set up the crank on a couple of good bearings in the block and slip in one rod and piston with no rings for the cylinder you are checking. You will need the cam you plan to use, installed and degreed as you plan to use it.

The heads should be assembled with one valve at a time using a light checking spring instead of the standard running spring. Now you can manually push on the valve to make it touch the piston for measuring clearance. With just one valve at a time you can turn the crank with finger tip pressure on the ratchet and feel anything that is touching or rubbing.

SCREW IN STUDS

The good guy Bowtie heads come with screw in studs. Mr. Average heads have pressed in rocker studs and before your new heavy duty springs jack them out, use a puller to remove them. Now you can have your favorite machine shop spot face the stud bosses, drill in the tap, and install new screw in studs, which is the right way to go. The job must be done on a mill.

You have a choice of 3/8 studs and standard rocker arms, or the heavy duty 7/16 screw in studs and beefier rocker arms. Both are 7/16ths at the bottom end. Part of installing screw in studs into early heads is to drill out the oval shape pushrod guide holes so they will not interfere. Guide plates retained by the screw in studs, keep the pushrod in alignment. The slots in which the pushrods ride should be polished and the corners smoothed out to avoid marring the pushrod surface.

When guide plates are installed, tighten the studs lightly, then tap the guide plate and position it so as to center the rockers over the valves, as you look at them from the exhaust port side. Now finish tightening the studs and recheck.

PUSHRODS

As the valve opens, the pushrods must stay straight in the face of considerable spring and acceleration forces. When a pushrod bends and resonates it adds to the rest of the valve train harmonics and can induce premature valve float, costing power and parts. Chevy heavy duty 0.075 pushrods work great. All cam makers offer good high grade chrome moly pushrods. Before you even put it into the engine, check the pushrod for straightness by rolling it on a flat surface. If it is bent and feels wiggly, it is trying to tell you something— it

Roller tipped rockers reduce the scrub action across the valve stem tip. Stud girdle ties in the studs so they support each other, which reduces flex.

bent because it was caught between the rock and the hard spot. Either the spring bottomed and coil bound or the valve spring retainer hit the guide, or the valve bent and hung up. Pushrods seldom just bend without a reason.

The pushrod should clear the block and head with room to spare. If you see rub marks on the pushrod, block or head it is time to do a little checking. With any sort of high lift cam check all pushrods for clearance through the valve travel. Keep in mind that the rocker arm brings the pushrod closer to

The pushrod guides are snugged down, but not tightened, to give you a chance to tap them in position and center the rocker arms over the valve stem tips. Studs are then fully tightened and the rocker position rechecked.

VALVE TRAIN GEOMETRY • 39

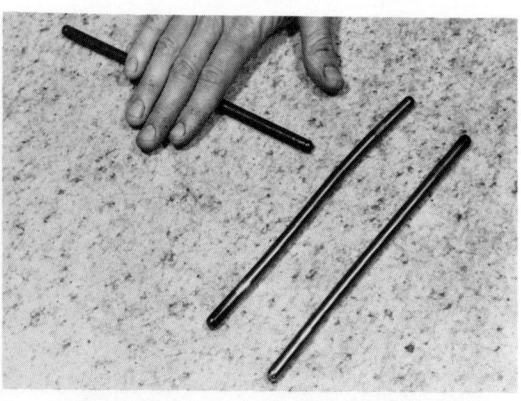

Pushrods must be straight—check by rolling them on a flat surface. Bent pushrods just don't happen. Mostly they are caused by spring coil-bind and valve spring retainers hitting the guides or by piston to valve collisions, which also spell bent valves.

the head at different points of its swing. One place most people neglect to check is interference between the pushrod and the rocker arm at full lift.

Pushrod length controls rocker arm geometry. As you go to a longer pushrod the rocker is raised and it will sweep the valve stem tip from a different angle—a point to which we will return in a moment. Pushrods are readily available in standard or plus 0.100 lengths and can also be special ordered to custom length. If you have a shop lathe, you can order pushrods with one end not pressed in. The tube is extra long so you can cut it to length and then press in the tip. Competition Cams has a neat pushrod length measuring fixture with two small Vee blocks for resting the pushrod, a stop at one

You can use a Moroso spring checker and read spring pressure, as installed, seated and open, on the torque arm scale.

end and a dial indicator at the other to insure consistency.

To find out the length of the pushrod you need, cut an old pushrod in two, and take a small slice out of the middle. Now drill and tap each half to insert a stud with a pair of lock nuts (some of the pushrods are ultra hard and cannot be tapped until they are heated and annealed).

When you raise the rocker pivot in relation to the valve stem, the tip of the rocker moves further toward the exhaust port side of the valve stem. You want to retain a well-centered sweep across the valve stem, not have the rocker come close to one edge and create damage. With a longer valve stem you are likely to need a longer pushrod so the rocker angle is retained. In a

Offset rocker arms on this Prototype prepared Jody Schmeisser engine gained extra intake port width.

more advanced engine, you may find yourself out of room and in need to move the rocker pivot, which in turn means moving the rocker stud away from the valve.

Think of a rocker arm as consisting of two lever arms, each with its own separate geometry. One arm extends forward from the rocker pivot to the valve stem and the other extends rearward from the rocker pivot to the pushrod. It is one rocker arm, but it consists of two individual levers, each with different lengths and angles. No matter what the number says on top of the rocker arm, the actual lever ratio is constantly changing through the valve lift and closing.

Since the rocker arm swings through an arc while the valve stem only moves in a straight line, constrained by the guide, little rolling action takes place

The Rimac gauge is the industry spring checking tool. Clamp on dial indicator is there for a height reading—spring pressure without a specific height is meaningless. Clamp on dial indicator is available from ABS.

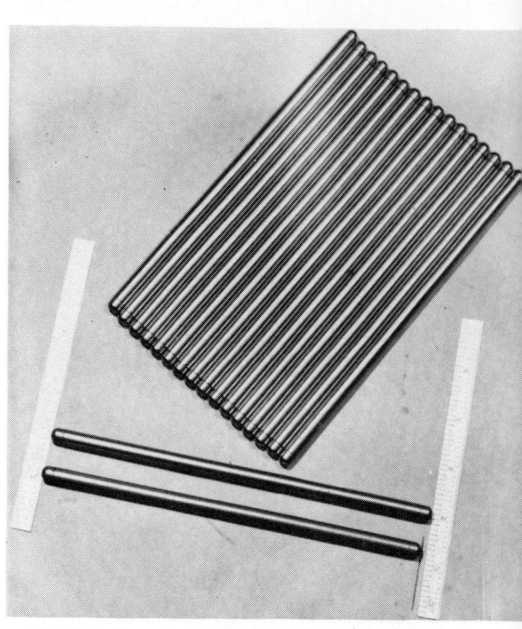

Pushrods are readily available in stock and plus 0.100 lengths. You can also get them with one tip removed, ready to cut to length, as a special order item.

between the two. In effect, the rocker arm is trying to not just push the valve up and down but also presses it on it sideways, a major contributing factor in guide wear. Smart racers set up the geometry of the arm to minimize that side motion. Point your left index finger straight ahead so it sticks out just like a valve stem in a guide, now set your right index finger to work like a rocker arm. The arc it describes can be nearly straight in relation to the "valve" or it can violently sweep in and out. You will need to center the sweep of the rocker arm tip near the middle of the valve stem. Spend the time to look at the

40 • VALVE TRAIN GEOMETRY

Check that the roller tips are centered over the valve stem. You may need to cut and re-weld the pushrod guide plates so they line up the rocker arms.

action between the rocker and the valve stem while you are turning over the engine and some of this hazy stuff we are speaking of suddenly sharpens up.

You are always fighting for valve train space; for instance, combining large springs and retainers with existing rocker arms creates clearance problems. With the valve seated, the valve spring retainer wants to nudge the inside of the rocker. Don't cure the problem by machining the rocker because that is where it will break. On the other hand, raising the rocker by going to longer valve stems and pushrods or to lashcaps can gain the necessary clearance. When you go to a higher number rocker arm ratio, which is usually gained by moving the pushrod closer to the rocker arm pivot, be particularly careful of the clearance between the pushrod and the rocker, at max lift.

In a good Pro/Stock motor there are all sorts of rocker arm tricks involved, such as moving the stud location back away from the valve. This gains pushrod clearance at the heads and block and it allows the rocker arm to center on longer valve stems.

SPRING RETAINER TO GUIDE CLEARANCE

Switching to a higher lift cam gains breathing and trophies, but if you don't do some prefitting it also earns more than a few headaches. Figure the valve spring installed height and the height of the spring at max lift. Now check that the spring will not coil bind at that height. Use a light checking spring, go to max lift and see that there is clearance between the valve stem oil seal and the retainer at maximum lift.

Work out your piston to valve clearances by setting up a dial indicator against the valve spring retainer and use light checking springs. Pushing on the valve spring retainer, in the vicinity of TDC, during overlap tells you if there is spare travel. You need a minimum of 0.075 on a tight high compression drag engine and 0.125 to 0.150 in a street engine, more on the exhaust than on the intake. With a low lift under 0.500, a casual check is good enough. As valve lift increases to 0.600 and up, considerable checking is needed. Beyond that you are coming closer to a pro motor and this section governs the entire engine.

OFFSET ROCKERS AND LIFTERS

As you go up in horsepower it pays to look into offset rockers that move the pushrods further apart. This allows more port width to be gained at a restrictive part of the port entrance. You'll need a weld build-up at the sides of the port, remachining to clear the new pushrod location, and now the ports can be widened. Another step in that direction is to use offset lifters which brings the bottom of the pushrods further apart for the same purpose.

ROLLER TIP ROCKERS

The best way to eliminate scrub between the tip of the rocker and the valve stem is to use roller rockers. They are a must on any decent race engine with a high lift cam. The roller helps protect the valve stem against wear and is often used in combination with hardened lash caps which give it a broader bearing area. Roller tips also reduce guide wear. Competition Cams makes an interesting steel roller tip rocker with a bathtub type pivot, inexpensive, light, and ultra strong. The next step up is a full roller rocker which has roller bearings at the center pivot and roller tips. They, too, are available in steel or in aluminum. If you use roller rockers, chances are the ones with the larger screw in studs will give you much less flex and longer life.

STUD GIRDLES

The forces involved in operating the valve are high enough to cause considerable flex at the studs. You can set up a dial indicator across the top of the stud, turn the crank and the swing of

When you are adjusting a hydraulic lifter turn the crank so the heel of the cam faces the lifter, snug down till the pushrod just turns, then turn in one more turn. That brings the plunger down from the top clip to gain extension travel. If you adjust the valves with manifold off, use the clip as a reference and move down one turn.

the indicator tells you the top of the stud is moving and that it is worth looking into stud girdles. The girdle simply ties in all the studs so that when one tries to move, the motion is resisted by all the adjacent studs. Stud girdles also help clamp the valve lash adjustment which eliminates having valves loosen up.

Valve train geometry is tricky business and your best bet, when you spot any problems during preassembly is call the tech people at your favorite cam grinder or consult with machine shop doing your work. You will find all info on recutting valve pockets in the piston section.

Rebushing the lifter bores of this Diamond Racing engine allows room for correcting block inaccuracy which is normally blamed on the camshaft. It gains the same timing in all cylinders.

VALVE TRAIN GEOMETRY • 41

COMPRESSION

You can cc a completed engine, or one being prefitted. This gives you a last minute chance to make compression ratio corrections with changes in head cc, gasket thickness, or block deck height. When Lee Bandrow cc's an engine during prefitting, he prevents leaks with a little grease to the ring land area of the piston after it has been brought up to TDC and uses a thin film of grease on both sides of the gasket and at the valves. This system gives you accurate compression volume cc.

An engine takes air and fuel into the cylinder, and then squeezes it to reduce its volume. That squeeze ratio is compression. Easy? Now let's add a few details. The piston is at the top of the stroke and we have some volume above it—combustion chamber, gasket and such:

THAT IS COMPRESSION VOLUME

As a piston moves from top to bottom it creates a volume and helps pull air +into the engine:

THAT IS SWEPT VOLUME

At the bottom of the stroke, BDC, the TOTAL volume above the piston is COMPRESSION VOLUME plus SWEPT VOLUME.

When a piston goes back up, we're back to compression volume. The ratio between the *total* and the compression volume is: **COMPRESSION RATIO.**

FIGURING COMPRESSION RATIOS

You buy a box of pistons that say 10.5:1, carefully install them in the engine, measure cc, and the answer could be anything from 9.7:1 onto 11:1.

It is not that the markings on the box are wrong. Somewhere in the catalog you will find a small asterisk that leads to small print at the bottom of the page. Here it says: "* with 66 cc chambers in a 355 engine."

With 74 cc chambers the compression is going to be lower and with a stroker 400 crank in a 350 block it will be higher. Time to figure your own compression ratios and do your own measuring and calculator button pushing.

Our forefathers saddled us with cubic inches, while modern folks who make burettes for measuring volume calibrate them in cubic centimeters. Convert cubic inches to cubic centimeters and do all the figuring in cc. The advantage is that if any changes need to be made, they can be measured with a burette in cc. Here are some handy conversion factors:

1.00 inch = 2.54 centimeters
1.00 cm. = 0.394 in.
1.00 cubic inch = 16.39 cubic centimeters (cc)
1.00 cc = 0.061 cu.in.

A plastic cover plate with a light coat of grease to seal it to the head gains an exact combustion chamber fill. Total cc include chamber volume, gasket volume, plus piston to deck and piston dish or dome volumes.

If you have a 350 small block, divide by 8 to get a 43.75 ci cylinder volume which is equal to 717 cc (43.75 X 16.39).

In any cylinder the volume is equal to base area times the height, or piston area times stroke.

If B is the bore and S is the stroke, V is the volume:

$$\frac{B \times B \times 3.1416 \times S}{4} = V$$

That is the volume for one cylinder.

For instance with a small block and a 4.00 bore and a 0.030 overbore you will have:

$$\frac{4.030 \times 4.030 \times 3.1416 \times 3.48}{4} = 44.39 \text{ c.i.}$$

Converting from cubic inches to cc for the rest of our example:

44.39 X 16.39 = 727.55 cc

The cylinder volume we have just gotten is called **SWEPT VOLUME or SV.** The volume above the piston at TDC is the **COMPRESSION VOLUME or CV.** It includes combustion chamber volume, gasket volume, the change made by a piston

42 • COMPRESSION

dome or dish and a thin slice of volume created by the piston being down in the hole or above the deck of the block.

Probably the world's easiest way to measure compression volume is to bring the piston to TDC, tip the engine on the stand so the spark plug hole faces up, and see how many cc it takes. During preassembly you can use the same system—the valves and the top of the piston around the ring land area are sealed off with grease.

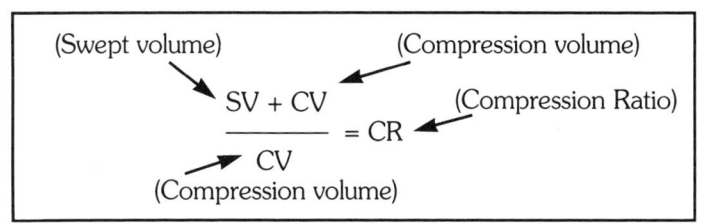

$$\frac{SV + CV}{CV} = CR$$

(Swept volume) → SV
(Compression volume) → CV
(Compression Ratio) → CR

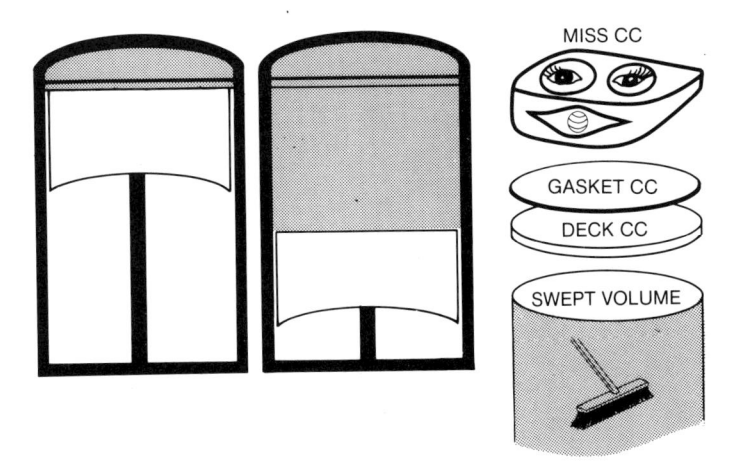

When you need to find the cc in the combustion chamber, first use this handy formula to find the compression volume—the volume above the piston at TDC.

$$\frac{SV}{CR-1} = CV \quad \text{(Compression volume)}$$

(Swept volume) → SV
(Compression Ratio minus 1) → CR−1

You are building a 9:1 motor from a 0.030 over small block. We have already figured out that it will have 727.54 cc swept volume. Since the compression ratio we are looking for is 9, we will have:

$$\frac{727.54}{9-1} = 90.94 \text{ cc or } 91 \text{ cc}$$

From this compression volume we have to deduct the gasket, deck and piston volume change (dish, flat or dome) to find the combustion chamber volume. As an example, figure 1.13 cc for each 0.010 that the piston is down in the hole and 9 cc for your Fel-Pro gasket plus 2 cc for the valve reliefs. Your chamber needs to be 91−13 = 78 with the piston down the hole by 0.020.

If you have a set of 74 cc heads with the same flat top pistons and a zero deck the compression volume you would be 74 + 9 + 2 = 85 cc compression volume. The compression ratio will be

$$\frac{727.54 + 85}{85} = 9.56$$

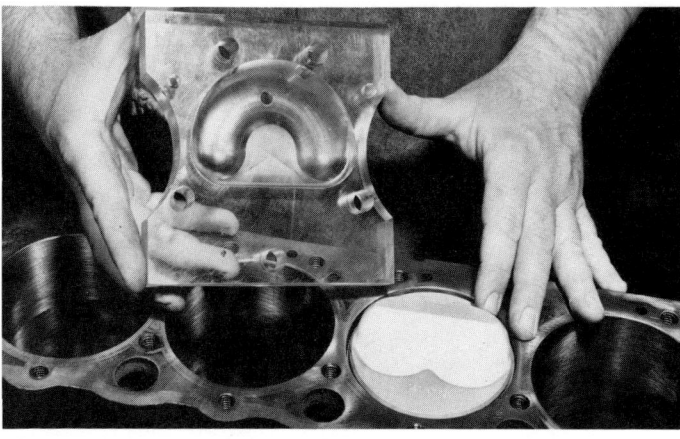

To measure piston dome cc, bring the piston down into the bore to a known depth and see how much fluid is required to fill up the remaining volume. Another way is to use this hollow plate, which clears the piston dome, seal it against the top deck and then check how many cc of fluid it takes to fill the known cavity volume. The difference between known and fill volumes gives the pop-up cc.

You can use a flat plexiglas plate to check cc in the head, and the same plate will also check cc within the block with a flat top piston. You can also used a dished plate to clear a piston dome.

A suitable method with a flat plate is to bring the piston down by a fixed amount, say an inch. You have created a volume the size of the cylinder, one inch deep. Calculate this volume and write it down. Now check the volume of fluid it took to fill up from a burette and also write it down. If the volume of fluid from the burette is larger than the calculated volume the difference between them is the volume of the dish in the piston. If the volume from the burette is smaller, the piston has a dome. The size of this dome is the difference in volumes.

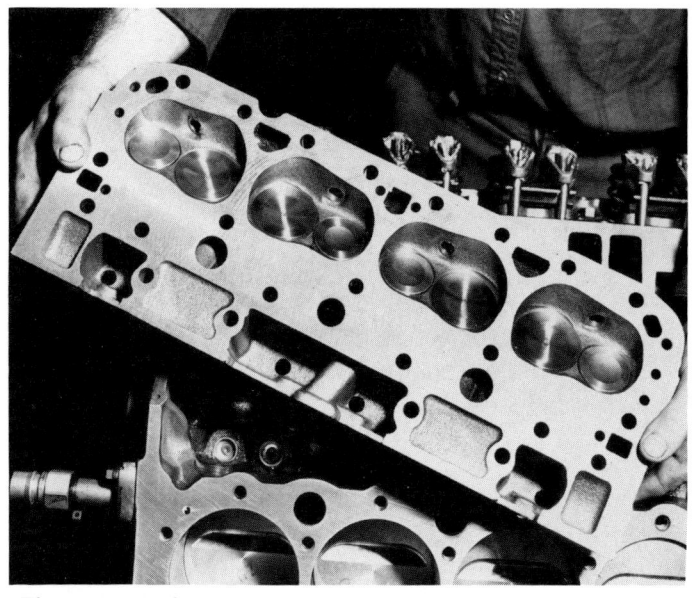

The amount of unshrouding you do around the valves increases the cc in the combustion chamber, but it also benefits engine breathing. High performance stainless steel valves are thicker than stock replacements and change the compression volume by as much as 3–4 cc. While the valve job is being done, check the heights and keep them all the same, because this affects cc.

CAM BEARINGS

Cam bearings are numbered: follow those numbers for location.

The oil drilling in the cam bearing must line up with the oil feed slot in the bore. Expanding cam bearing drivers are common, but do not provide good bearing support, compared to a custom fitted one. It is good practice to install Chevy cam bearings with the oil feed holes at approximately four o'clock when viewed from the front of the engine, with the oil pan down. This reduces oil leakage between the cam and the bearing.

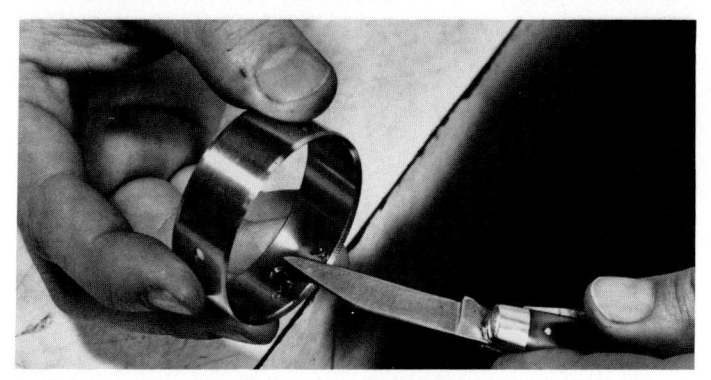

A light scrape around the edge of the cam bearing makes it easier to insert the cam.

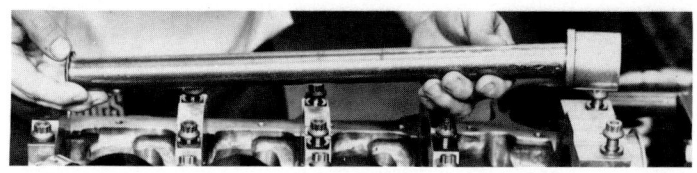

The best driver has a shouldered pilot which holds bearing size and gives it full support—make your own on a lathe.

In a Chevy, all of the oil to the mains is first routed through grooves around the cam bearings, and from there to the mains. When the block is professionally degreased, the cam bearings are removed to gain access to the grooves and galleries behind the bearings for a better cleanout. If the bearings were left in, the clean out chemicals would dissolve what's left of the bearing material. A bearing driver with a machined shoulder and a screw-in long rod is used to hammer out the bearings. The plug in the block at the rear bearing is removed at the same time as the bearings.

After the bare block has been degreased, the machine shop will debur the cam bearing bores and install new ones. The deburring is usually done with a long rod chucked in a drill. A slot in the end of the rod holds medium grit flapper paper. This eliminates casting burrs, any traces of sealing material, or burrs left by the driving tool. Checking the cam bearing bores with a small dial bore gauge helps prevent bearing installation problems. The more sophisticated race shops have a cam bearing line bore set up, but that is the high end of the business. Your best bet is to ask the machine shop doing the block work for you to install the cam bearings and fit them to the cam. However, you should also know the procedure and have some input, to avoid common errors.

Most engine shops install cam bearings with a universal tool. It consists of a rubber support center and a drive shoulder. A number of the better engine shops machine their own drive tools. Here the pilot for the bearing is of the same size as the cam journal plus an extra couple of thousandths for bearing clearance. The solid tool also has a shoulder for driving the bearing. This tool offers a better pilot action and keeps the bearing from caving in. Cam bearings are numbered and vary from journal to journal. The number 1 bearing is the largest, numbers 2 and 5 are the same but smaller and number 3 and 4 are the same and still smaller. We are speaking of the bearing's outside diameter. The cam journal sizes are all the same.

If what follows sounds a little bit like the Stone Age we quite agree. Production Chevy cam bearing bores do not appear to be line bored originally. Semi-finished bearings are installed and are line bored on a production machine line so the cam turns true and with the right clearance. The bearings are not heavily loaded and last a long time. However, when you install new bearings, the cam may go in tight. Now a little machinist's blue is applied and

44 • CAM BEARINGS

wherever shiny spots show up a bearing scraper is used.

Just in case you are not familiar with a bearing scraper, it has a small handle, a triangular cross section, with a point at the end and straight sides—it looks like a hollow ground three corner file with smooth sides. In fact, you can make your own from one. The edges are extremely sharp and are great for scraping a fine amount from the bearing surface. It will only stay sharp if you hone it and avoid scraping too much steel with it.

Smart race engine builders don't put in all the cam bearings at once—they install a pair of bearings two or three saddles apart, try the cam and when it fits nicely, they keep adding bearings one at a time and rechecking. Cam bearings are a press fit. Once you drive them in they are swaged down by the bore size. If you have to take them out again to refit keep it or fit a new one, but in either case, use an anaerobic sealer (Red Pro/Lock or Loctite). One inexpensive bit of insurance is to pretest the bearings by slipping them over the cam. Also a very slight chamfer made with a bearing scraper helps lead the cam into the bearing bore.

Here are some handy bearing numbers to keep in mind. A Sealed Power SH-290S has a babbitted lining on a full round bearing housing—not split. The SH-1090S has a tri-metal lining and a split outer housing. Also available is another very handy number SH-1196S which is an oversized bearing good for salvaging a block where a cam bearing has spun. It is particularly good for line boring the cam saddles in the block.

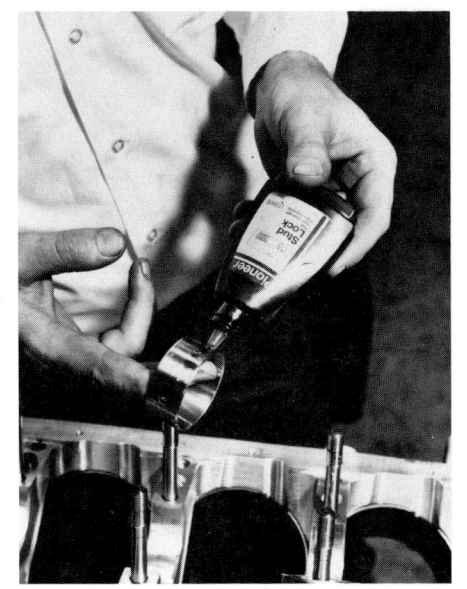

An anaerobic sealer on the OD of the cam bearing gains a tighter grip.

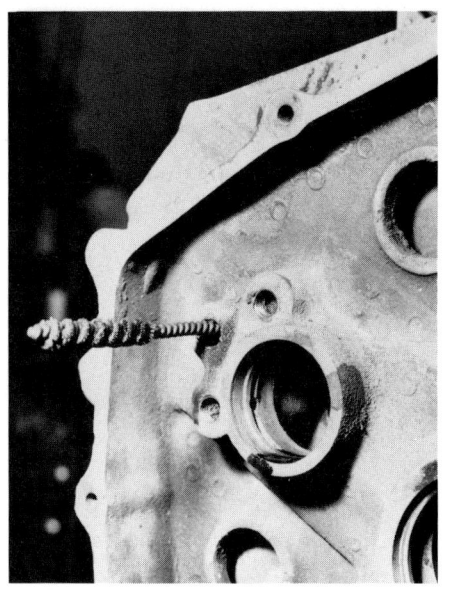

Groove and oil gallery must be cleaned before installing the cam bearing. Bearing oil supply hole must line up with the groove.

Now you can change cam bearings fast and know that they will be in line. This is very popular among NASCAR Winston Cup racers and some top drag race engine builders use it.

Bill Howell gave us some interesting insight on the cam bearing oil hole location and the reasoning behind it. You would normally think that since there is a full oil groove behind the bearing shell it would make no difference. Bill explains that the cam sees mostly a downward load from the lifters. If the oil hole is placed at the top, say at 12 o'clock, looking at the cam from the front of the engine, with the oil pan down, the oil will be delivered in the largest clearance area, and this can account for a pressure loss.

Keep in mind that the cam turns clockwise, again looking at it from the front, which means that it will ride on a wedge of oil. Best oil hole location is a bit down from center, say at 4 o'clock. The clearance is moderate, and the cam entrains the oil for max lubrication. If you placed the oil hole at the bottom, the cam would tend to shut off the oil, and if you placed it at 9 o'clock the pressure would dissipate itself along the top and never reach the bottom. Some prefer to offset the oil hole slightly to limit cam oil, but if you place it at the correct angle it tends to self limit, and you can simply line up the hole with the groove.

The cam bearing bores in the block should be flapper papered to remove rough entry edges or varnish deposits. Any time the block is boiled out, the cam bearings must be out to insure adequate cleaning.

Install a couple of cam bearings at a time and check that the cam goes in and turns freely. If need be, the bearing must be carefully scraped at the high spots.

CAM BEARINGS • 45

INSTALLING A CRANK

A dial bore gauge is the most accurate way of measuring the inside of a bore to the nearest tenth (0.0001 or tenth of a thousandth).

The crank is first laid in without a rear main seal and with the bearings dry, for Plastigauge checking. If all the bearing clearances are within limits, take the crank back out, oil the bearings, install and tighten the caps, and see if it turns freely. This spots a bent crank or main bearing misalignment in the block. Now install the seal, final assemble the crank, torque the bearings, and recheck crank turning action.

It is final assembly time and you are ready to lay in the crank. Unlike a wedding you'll enjoy a final chance to find out if the bearings and the crank were really meant for each other. First step is to see if the crank is straight. Put two clean used main bearing shells in the front and rear mains, oil them up and lay in the crank. Then set up a dial indicator on the No. 3 main, zero it, and check for runout. If you want to be careful, check the nose of the crank separately to indicate runout.

At crank shops, cranks are stored in vertical racks to keep them straight and also to get them out of the way. Park your crank so that it does not get bumped into and accidentally dumped on the floor. To check crank runout place it on two bearings and use a dial indicator: 0.001 to 0.002 is the generally accepted figure.

Take a fine file and gently break any sharp corners at the parting faces so that they will not scrape off pieces of the bearing on installation. Any time you insert a bearing, set it in tang side first and use some thumb pressure to slide in the other end of the shell. If you have to remove the shell, grind a nice rounded end on a small screwdriver and use it to tap on the end opposite the tang—the bearing will roll right out.

Check the alignment of the oil delivery galleries in the block with those of the bearing in the grooved upper shells. If need be, correct any misalignment. Next, test the bearing shells against the crank to see if they need machining at the ends to clear the radius areas. Set the shells in the block and in the caps, to see that the tangs are not pushing away the bearings. If need be, do a little filing on the bearings. Interference at the tang causes the bearing to heat and wear excessively.

Check the numbering, the fit of the cap in the block, and the alignment of the cap fore and aft. The parting face edges of the cap and block should match and this may not be perfect on production blocks. Also, the machining or cross-hatch patterns should match across the parting faces. This check prevents mismatched caps and blocks. With the crank out of the way and the bearing shells in place on the block and caps, but with no crank, check that in any one bearing saddle neither bearing shell is ahead of the other one, or they

You can open up the top bearing shell feed hole so it lines up with the block oil gallery and doesn't restrict flow.

46 • INSTALLING A CRANK

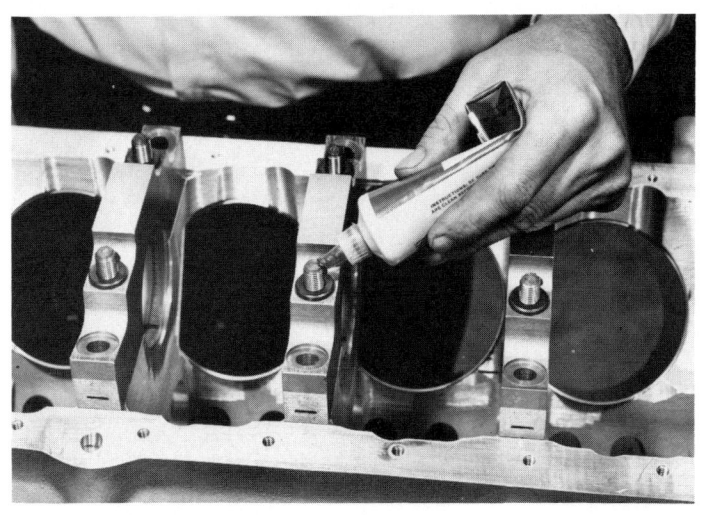

Use studs and hardened washers at the mains and lube both the threads and the washer.

The black rubber seal has a single lip while the gray one uses dual lips, one to keep out dirt and the other to retain oil. Spiral grooves help the crank pump back oil to the sealing lip. Make sure the crank sealing surface is ridge-free.

Lay down a strip of Sealed Power's Plastigauge to check bearing clearance.

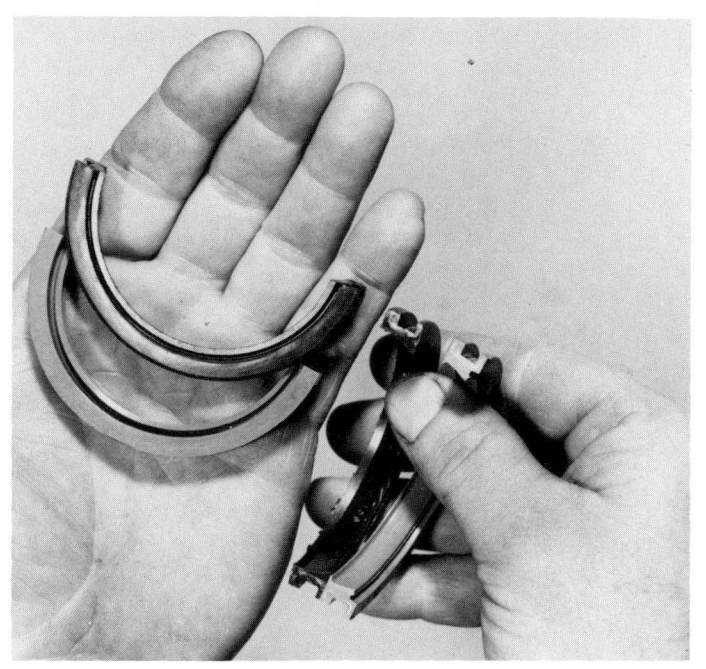

The thin thread-like Plastigauge material is squashed by the bearing, and you can read the bearing clearance by checking its width with the wrapper markings.

begin to act as scrapers and starve the other one for oil. If need be, indicate with stake marks where the main bearing saddle should fit on the block lengthwise, because once the crank is in, it is hard to check bearing alignment.

When you install main bearing caps, do not draw them down with the bolts until the cap is seated at the parting face. That important locating area is an interference fit and you don't want to mangle it by drawing down with bolts. Threads on main bearing bolts need a coat of oil and so does the underside of the bolt head. If you convert to studs, use Pro Lock II at the coarse threads, oil is used at the threads and the nut. Bolt torques in manufacturer's specs are given with motor oil and more slippery materials call for less torque.

The coarse part of the stud, coated with a sealer, is just lightly tightened and allowed to set up. If you bear down on it with full force, the unthreaded portion wedges in and splits the casting. This hold for all studs.

There are two ways of checking bearing clearances. The most accurate is to use a dial bore gauge to first check the main bearing saddles in the block. The caps are bolted to the block without bearings and the saddles should be within tolerance range. Next, check the inside diameter of the bearing shells installed in a block, with the caps torqued down. Your bearing manual will have all the figures. You can get a bearing manual from many sources such as Vanderwell, Speed-Pro, Federal Mogul, or Michigan.

The difference between the inside diameter of the bearing (bearing ID) and the outside diameter of the journal (journal OD) is the bearing clearance. If you do not have what you want, improve things with a little mix and match work. You can intermix standard and one under journals to cut down on unwanted clearance. A ball or tubing mike can be used to check the individual shells to grade them.

As a final check, install the bearings in the saddles and the caps dry. Set in the crank and put a strip of Plastigage along the crank at each bearing saddle. Check one bearing at a time and you will find the crank clearance. This last

INSTALLING A CRANK • 47

step, when you have already measured everything with a dial bore gauge is redundant, but waves off measurement errors.

If you don't have a dial bore gauge or mikes, then the Plastigage becomes the main size check. Say the mains are between 0.002 and 0.0025 inch and you don't need further fine tuning. Remove the caps and the crank, clean everything up one last time, coat the bearings and the crank with GM EOS, or your favorite assembly lube, install one bearing cap at a time, line it up to the stake marks, tighten, and check how freely the crank spins. Keep on going all the way back. Don't forget to install the oil seal first if it is a two piece lip seal.

occasionally from a torque converter is directed forward (converters can expand lengthwise under pressure) concentrate on lining up the rear thrust faces. Some builders find it helpful to push the crank all the way forward and wedge it. You can then use a mallet to tap the cap rearward and seat it against the crank with both rear thrust faces aligned.

To check crank end clearance, set up a dial indicator so the pointer is in line with the axis of the crank. Bring the crank all the way back with a screwdriver or a pry bar, zero the indicator and then move the crank forward again. This gives you the end clearance.

If you do not have enough end clearance, first try again. Should this still

crank and put everything back together.

If you have a 400 block that has been line honed, the rear main seal saddle has been cut. Now you need to make up an aluminum shim—some builders simply use a tie down strap—to place the seal at the right height. There should be a *light* coat of sealer applied to the inside groove at the back of the seal. You will also need a *light* amount of sealer at the back end of the block and the main bearing cap. Any sealer on the contact face of the seal itself is damaging and needs to be wiped off. Also, any excess sealer should be wiped off so it does not wind up on the oil pump screen.

Bob Gillelan of Moldex Tools points out that there are two very common

Use engine assembly lube on the inside of the bearings. Back of the bearing goes in dry.

After the bearings are locked up, check the crank end clearance with an indicator. Pry the crank front and rear to seat the main bearing cap, and tighten. Zero the indicator with the crank all the way back, pry the crank forward and read end clearance. Torque down and recheck.

END CLEARANCE

Crank end clearance is controlled by the rear crank journal and by the rear main. The first quick rough check is to lay the rear main bearing shells on the crank one at a time and measure end clearance with a feeler gauge. At the same time inspect the seal area and the crank thrust faces on the crank.

The thrust faces of the upper and lower main bearing shells work together as a team, so you don't want one ahead or behind the other. There is a certain amount of clearance available at the main bearing cap bolts so that the cap can move forward or back to line up the top and bottom thrust surfaces. With the cap tightened slightly, the crank is moved forward and then back which moves the cap into alignment. The main is then locked down. Since most of the thrust from a clutch and

does not work, remove the rear cap and recheck. If you now have the correct clearance, the problem is in the cap location or in the cap itself. When the locating dowels are the problem, a little filing will help. A cap can throw you off if it is not cut correctly.

If the caps were touched up at the parting face for line honing, it is entirely possible for the cap to get cut out of square. Any dirt or grinding debris on the surface grinding fixture that locates the cap, can set it at an angle and cause it to be cut that way. To find out, use a little Dykem on the thrust faces of the bearing in the cap and see where they contact. If you only find contact at the top front of the bearing, you know the cap is tipped and a mini touchup at the bearing will regain the clearance. With the right end clearance and confidence that the crank spins freely, remove the main bearing cap and the

sources of damage to the crank nose. One of them is oversized, anodized keys at the timing gear and damper. If they are hard to drive in, don't use a bigger hammer. Put a piece of emery cloth on a flat surface and take down the key so it is just a snug tap fit. Problem is that an oversized key puts a huge stress on the machined corner in the crank keyway and that is where the crack originates—it is just like splitting logs.

Another damage source are off-breed timing gears with not enough inside bevel to clear the snout radius at the crank nose. When the timing gear is driven on with a big hammer it forms a ridge in the radius and the snout breaks off, obviously a bad crank, since it broke. If the bevel does not look right, or if the gear is not going on flush against the crank, don't hammer, check the bevel on the inside of the gear.

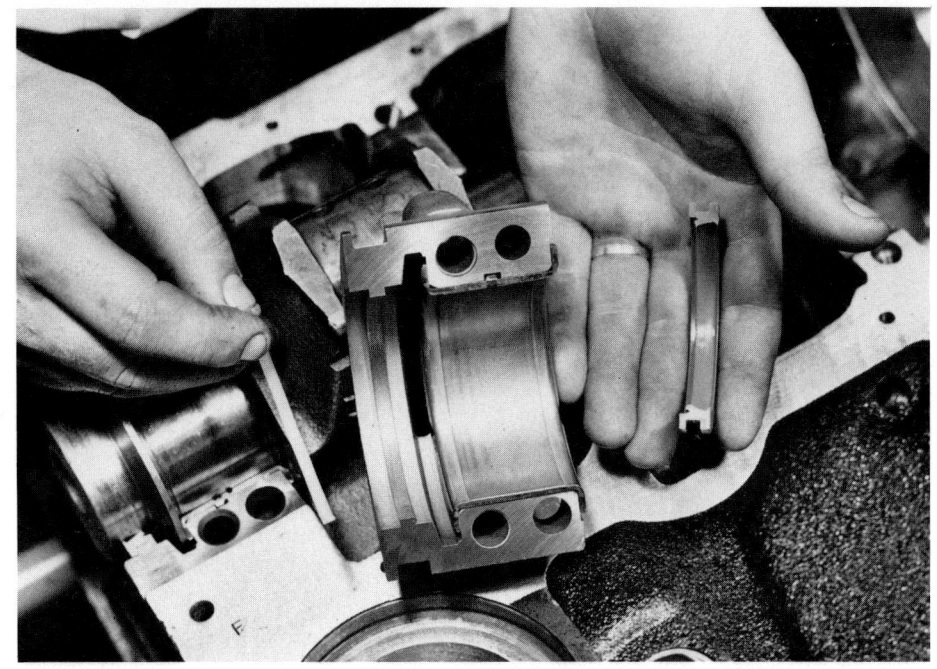

To line hone a 400 cube engine, the rear seal support lip must first be bored out to main bearing size. To restore the seal to a correct height, fabricate an aluminum or brass shim or use a tie down strap cut to size and insert between the seal saddle and the seal. Raised too high, a seal will burn up. Not raised high enough, it leaks.

A dab of sealer on the ends of the rear seal and a film of sealer along the rear surface and corner will prevent leaks. If you don't have enough end clearance, remove the cap and see if the clearance increases. This would tell you the cap needs to move or may be out of square.

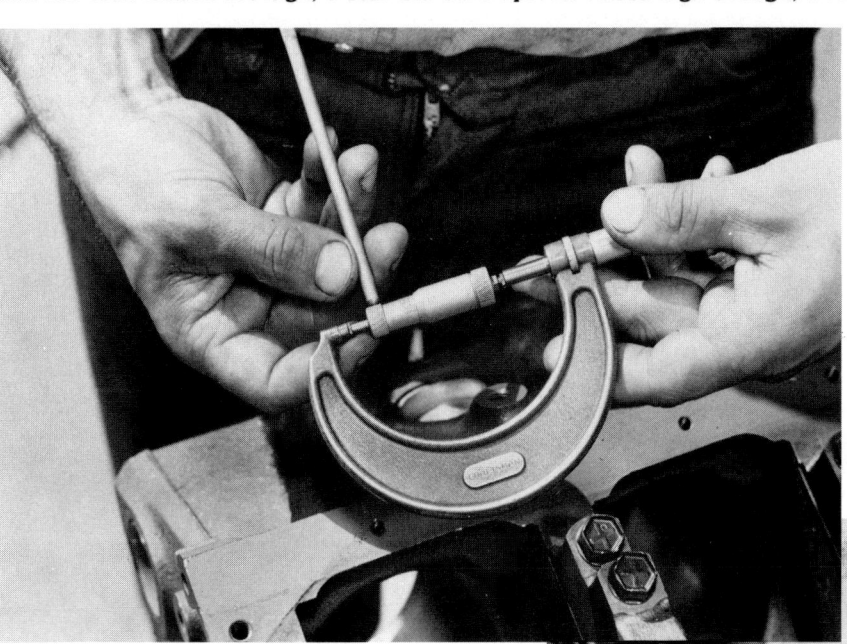

To check bearing sizes you need a mike with a ball anvil—a rounded surface that clears the inside curve. Bearings in a set are not all equal and miking them helps grade the bearing shells, for a better match to the individual journals. You need a mike that reads in tenths of a thousandth—with a vernier read out or a digital mike.

Yes, a dial indicator is definitely best, but the affordable inside mike lets you size the inside of a bearing bore or a bearing shell. Once you get the hang of it, it is quite accurate. Set the inside mike so it rocks smoothly in the bore, lock up with the handle and now take your readings with the same mike that you will use to check the crank journals. The clearance is the difference in readings. The advantage is that you have eliminated two reading errors and are getting a direct difference value.

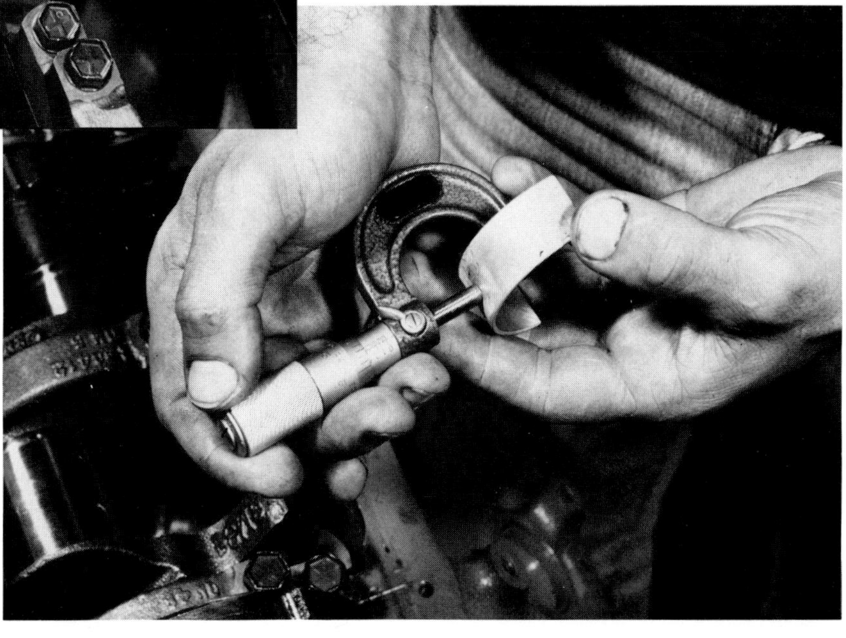

INSTALLING A CRANK • 49

DEGREEING A CAM

Moroso offers two choices of degree wheels: big and giant. To analyze cam profiles, get the big one—it gives fine readings. For normal degreeing, the small one is more convenient.

There is no mystery or magic in degreeing a cam—it just helps you put the cam where you want it in relation to the crank. It is also a handy tuning device; for instance, you can use degreeing to gain or lose torque at the starting line, or to trade low rpm torque for extra pull at the top end.

Moroso's engine builder and dyno expert, Bob Rinaldi tells us you can definitely feel a two degree change in cam setting either at the track or at the dyno, allowing that the engine is well tuned. Unfortunately, stock timing marks are not reliable.

Main reason for degreeing a cam is a pile up of small individual machining errors which may either cancel or add up. Keyways in the crank or crank gear may be off. Some crank gears with multiple keyways are notoriously inac-

Before you fully coat the cam with assembly lube, first try it out in the bearings to see that it can turn freely. Now coat it for final assembly.

The degree wheel comes with three different bushing sizes, which adapts it to the crank bolt.

curate, others are excellent. More errors are added at the cam pin and at the bore in the gear. We have seen OEM gears with timing marks designed to retard the cam for emissions reasons, and cam pins with substantial advance.

Degreeing a cam begins with locating Top Dead Center (TDC) and then phasing the cam so the valves open at the correct time in crank degrees. Degreeing also means advancing or retarding a cam to suit torque requirements or checking the lobes in a cam to see if they are accurate. You can spend 15 minutes or half a day depending on how fussy you want to be.

Smart race shops save considerable time with the camshaft analyzer built by Quadrant Scientific in Louisville, CO. This unit allows you to first check all lobes on the cam itself, giving their full profile and location and to then again check the same cam in the block. This can reveal on a printout in a fraction of the time what you need to know about the cam and block and also allows you to get much deeper into seeing what is in a cam that happens to work better for you. It is both a quality control tool and a learning tool to "see" the design features of a cam.

A cam needs a little end play so that oil can penetrate at the thrust button and between cam and block.

DEGREE WHEEL TIPS

We want to measure the crank position in degrees: naturally, the larger the degree wheel, the easier it is to accu-

Always make sure that the damper timing marks correspond to TDC or modify the pointers to match. If the damper is way off, it can no longer hold timing because the rubber allows the ring to slip.

50 • DEGREEING A CAM

rately read the marks. For instance, Moroso offers two convenient sizes—an 11 inch wheel handy for use in the car or on an engine stand and an 18 inch wheel which makes it easier to check cam profiles and makes it more of an assembler's tool. Always check that there is no runout at the degree wheel when it is mounted on the crank. If the wheel is installed off-center, you can end up measuring 178 degrees on one side and 182 degrees on the other, which defeats the accuracy of the cam installation. Bushings are supplied to center the degree wheel on different crank bolts.

Moroso's handy combination crank turning socket, and degree wheel

Check out the damper timing markings as part of degreeing a cam. This Moroso insert protects the crank threads from the puller.

Use a dial indicator at the valve or at the cam. Either way it must be set up so the pointer moves in the line of motion—not at an angle.

This handy dial indicator helps you find TDC and can also measure the piston to deck clearance. Because the piston has side clearance it swings a little inside the bore. You will only get consistent TDC readings when the dial indicator side of the piston is pushed up.

mount, is worth its weight in gold. A half inch drive socket, built into one end, accepts a ratchet or a breaker bar. The socket slips over the crank nose, engages the damper drive key and turns the crank. You can tighten the set screw in the drive socket and eliminate all play. The degree wheel slips over the threaded extension on the end of the crank socket and is locked in place with a knurled knob. It is a lot easier than bending pointers or having a degree wheel bolted to the damper.

There are 360 degrees in a circle but instead of numbering all around, the degree wheel runs from zero at TDC to 180 degrees at BDC and then goes from zero to 180 degrees on its way from bottom to top. A set of smaller inside numbers read in reverse. This is handy for exhaust opening is given in degrees before BDC and intake opening is read before TDC. Conveniently marked by color changes in the degree wheel areas are ranges where intake and exhaust opening should occur. Rare would be the occasion where you would have a cam outside of those ranges, which blocks out gross errors. Also marked, are the ranges for lobe center lines (max cam lift points).

POINTERS

A handy pointer can be fabricated from a welding rod, or from a piece of steel wire, to mount at the water pump outlet boss on the front of the block.

Keep the pointer close to the degree wheel, make it rigid, sharpen it to a wedge, and take your readings straight from the pointer to the marks. Looking from the side can introduce errors in the reading.

FINDING TDC

When you bring the piston all the way to the top, the crank is able to swing several degrees either way without affecting the piston motion. In other words, just bringing the piston to the top is too inaccurate in finding TDC. One way is to make a piston stop—a simple crossbar with a center bolt. The bar is bolted to the top of the block and now the piston comes up against the stop on either side of TDC. Read the two values and move either the pointer or the degree wheel, to make them come out equal. You can also use a Moroso "bridge" with a dial indicator over the cylinder bore.

MARKING THE DAMPER

Most dampers have rubber between the hub and the outer ring. When the rubber ages and is overstretched, the ring slips and the timing marks become useless. Sharp racers put two punch marks on the timing damper—one on the ring and the other on the hub, exactly opposite each other. If the ring

When you use a puller, thread the bolts far enough into the damper to get a good bite and keep the puller even with the damper. Do not hammer on the damper to drive it on—use a threaded installation tool with a thrust bearing.

DEGREEING A CAM • 51

A Torrington thrust assembly protects the front of the block and reduces drag. Machine either the front of the block or the inside of the cam sprocket to provide the extra room, and to compensate for misalignment between cam and crank sprockets.

TIMING CARD NUMBERS

The first time you look at a timing card, it could have just as well been written in Chinese. If you want to make some sense out of one, here is a short list of all the markings you will see:

TDC Top Dead Center or top of piston travel
BDC Bottom Dead Center or Bottom of piston travel
BTDC Before Top Dead Center
BBDC Before Bottom Dead Center
ATDC After Top Dead Center
ABDC After Bottom Dead Center

The timing card will also use abbreviations for intake opening and closing and exhaust opening and closing:
IO Intake Opening
IC Intake Closing
Same for EO and EC For instance, IO 37 degrees BTDC is short for intake opening 37 degrees before top dead center.

IT'S BACKWARDS

One of the little things you have to watch out for is that the degree intake valve opening diagram on the timing card is on the **LEFT** side of the diagram, and shows that the opening occurs before TDC. The degree wheel has the opening occurring on the **RIGHT** side of TDC in a **RIGHT ROTATION ENGINE** Obviously, the mark at which the intake valve begins to open is going to reach the pointer before the TDC mark on the degree wheel. You just have to mind-flip the timing card marks to match degree wheel marks.

DEGREEING THE CAM

Now that the degree wheel is set up at the nose of the crank, complete with pointer, use it to check cam location in relation to the crank. You can also find the profile of each cam lobe, and whether the cam lobes are phased correctly with each other. You read the degrees and check lift. This enables you to plot a cam curve and keep a record of what you have.

For the most part a cam coming from a modern manufacturer will have all the lobes in the right place and you only need to check one cylinder.

Winston Cup teams with mega bucks are known to check every lobe on electronic test equipment, and to send back cams that don't pass muster. They also check lifter bores and redo them. It all depends on how far you want to go.

When the engine is apart, with just a cam and crank you can accurately degree the cam from the lifter. When the engine is together, it is easier to degree from the valve spring retainer. The two measurements will usually not agree because cam timing varies with deflections introduced in the valve train geometry. You may also find variations from cylinder to cylinder, that are due not to cam error but to the fact that the lifter bores are out of position.

When checking at the valve spring retainer be sure to use a solid lifter or a blocked up hydraulic lifter. A standard hydraulic lifter leaks down and the values change.

DIAL INDICATOR

Most dial indicators have a magnetic base and will fit on top of the block with no problem. To degree from the head, bolt on a steel plate like the one that Moroso supplies, from the valve cover rail to the head. It is useful for both cast iron and aluminum heads.

You need an indicator with a one inch travel which keeps you from running out of room with higher lift cams. Time is saved when working with two indicators, one at each valve. An indicator must be set up on a firm base with no shake in the linkage. Also, the rod on the indicator needs to be lined up with the motion of the lifter or spring retainer. Set at an angle, it cannot give a true reading.

An indicator is a comparator—it can measure from any zero that you set, but if you don't set it to zero, it does not know what to compare to. Don't use the first zero that you set: instead go through a couple of crank turns to make sure that the indicator returns to the same zero. This will often not be the case until you eliminate lifter sticking and wobbly supports.

BASE CIRCLE

Each lobe has a base circle or low lift area. Ideally, there should be no runout and that is a good zero for setting the indicator. Actually, some runout is inevitable. Let's turn the crank and see what happens at the lifter. It will go from zero, to a very slow lift in the initial ramp and then a faster lift along the flanks. Eventually, the lifter has to slow

Push the cam and crank sprockets back toward the block and check alignment. Compensate for misalignment by machining the cam sprocket at the cam face.

With an aluminum head, make up a rigid steel adapter plate to serve as an indicator stand.

slips, you can spot it right away and discard the damper.

Once you have found TDC, use the degrees wheel to help locate spark advance marks from 40 to 30 degrees. This allows you to check total timing. You can also add a Moroso tape to the damper. Wipe the damper with thinner to get better adhesion.

52 • DEGREEING A CAM

Use a thrust button and a retaining plate. If you don't bend up the lock tabs, they won't break. Also, don't do as in this picture—use new Grade-8 bolts and lock them with an anaerobic sealer.

Always turn the engine to the right, as it runs, so as not to induce slack in the drive side of the chain, find maximum lift, and then keep turning till you come to 0.010 before max lift and 0.010 after that maximum lift and note the degree readings. Split the difference in degrees and that is your max lift point.

down, come to a full stop, and reverse direction. You can then turn the crank through a full turn showing no motion, and the lifter begins to move again.

Cam opening points are very gradual. That slow initial ramp helps take up all the clearances on a solid lifter cam to apply the load to the valve train smoothly and evenly. The closing ramp controls the lifter return and the smoothness of the valve closing. Hydraulic lifter cams also have ramps, but much shorter ones since they run at zero lash clearance. Ramps make it difficult to find accurate opening or closing degrees. It is much easier to measure timing at a 0.050 or a 0.100 lift. Here the lift takes place at a higher rate, the lobe flank is steeper and it is easier to pinpoint the timing. The cam maker specifies on his timing card how you should measure opening and closing points on his cam.

CAM CENTER LINE

The most convenient timing method for a performance cam is to degree not from the opening and closing points, but rather from the cam center line. By definition, the cam center line is at the maximum lift point. Center line is a bit of a misnomer, a carry over from the days when cams were all symmetrical. Most modern cams are asymmetrical and have a faster opening and a slower closing, so what we call a center line is just a maximum lift point.

Finding the maximum lift point with a dial indicator is easy but does not give you a center line directly because the cam nose is very radiused. Instead, you wind up with a dwell area of several degrees where the degree wheel is moving and the lifter is standing still, or moving very little.

The reliable way is to first find a maximum lift figure. Set your indicator to zero, back down to 0.040 on one side of maximum lift, and read the degree wheel. Now swing to the other side of maximum lift and again read the degree wheel. The mid-point between those two degree readings is the center line.

The best way is to always turn the crank continuously in one direction as to maintain tension on the timing chain, which involves some extra turning. The other way is to back up past the desired mark by a quarter turn and then turn forward to take up slack, and get your readings. Once you have located the first center line figure, take a couple of insurance policies by rechecking at 0.030 and 0.020 below the max lift. The figures will be very slightly different, but right in the ballpark.

On a cam with modest duration and lift, the nose of the cam is flatter, and finding the center line is not very reliable. You are better off to work with opening and closing points at 0.050 or 0.100 lift.

Moroso has a crank turning socket which allows you to mount the degree wheel. Manufacture your own pointer from flat stock or welding rod and sharpen the tip.

DEGREEING A CAM • 53

RODS RINGS PISTONS

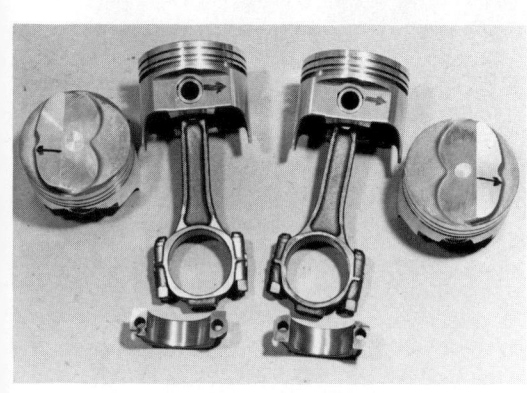

There are front arrows on the pistons, and the spark plug sides face out. Also the bearing tangs on the cap and the rod face out. The long side on the piston goes in the direction of rotation, as you face the front of the engine.

The block is together, the crank is in and you have already matched and Plastigaged the bearings. Our next moves are to hang the pistons on the rods, install the rings, and to assemble the block. It's easy to do and even easier to blunder into all kinds of mistakes. Keith Burton from Jeg's in Columbus, OH took time out to let us shoot all the pictures and to provide an infinity of

The rings have been prefitted and you are ready to install them. Combine one stack and one piston at a time.

detail. The same also took place at LAB Machine in Lindenhurst, NY and at Seaport Automotive in Toledo, OH. The rings have been prefitted and are arranged in numbered stacks, one set for each piston. We all know not to mix the rods and caps, but you have just had three phone calls, two interruptions, and finally locked the door. Each rod was reconditioned, fitted with new bolts, and honed out. The cross hatch marks must line up from the cap to the rod. Also, if the cap is mismatched you will feel and see a misalignment between the cap and the rod.

As a rule of thumb, both bearing tangs face each other on the same side. In a Chevy the tangs on the rod face out—the outside of the block—point to the right on the right side and to the left on the left side. Also, if there is a bevel in the big end of the rod, the bevel goes towards the radius of the crank.

Here are a few small pre-checking points. Check the bearings against the crank, just to see that the box was not mislabeled. Take one pair of rod caps and set them against the crank, as a fast check of rod width. Take a compression ring and spin it around each piston groove, from the outside, without ever installing it, just to see that the grooves are clear.

HANGING PISTONS ON THE RODS

The panic is on—which way did the pistons go? Where did I put those notes? Easy!! The piston usually has an arrow facing forward. Piston pop-ups face toward the outside. On the higher compression flat top pistons, the valve pockets face the intake side. Racing pistons will normally have the wrist pin in the center. For quieter running, the street piston pin uses an offset pin; when you face the engine from the

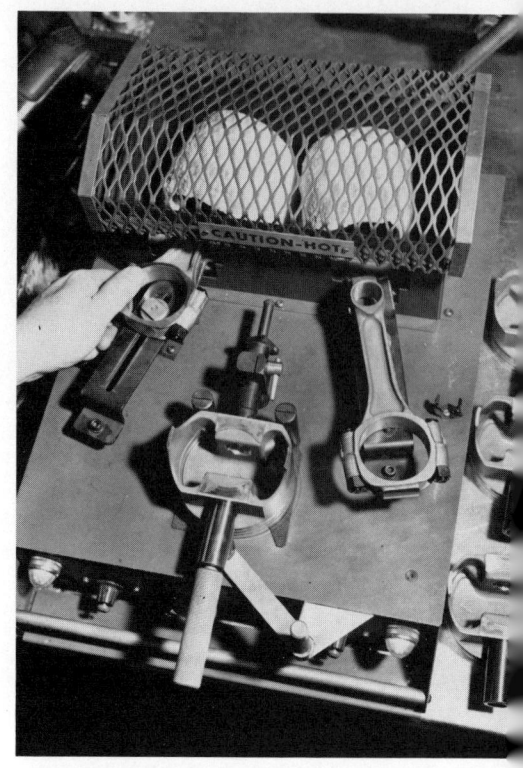

The rods are heated, to expand them, and the piston is in place with the pin and tool ready.

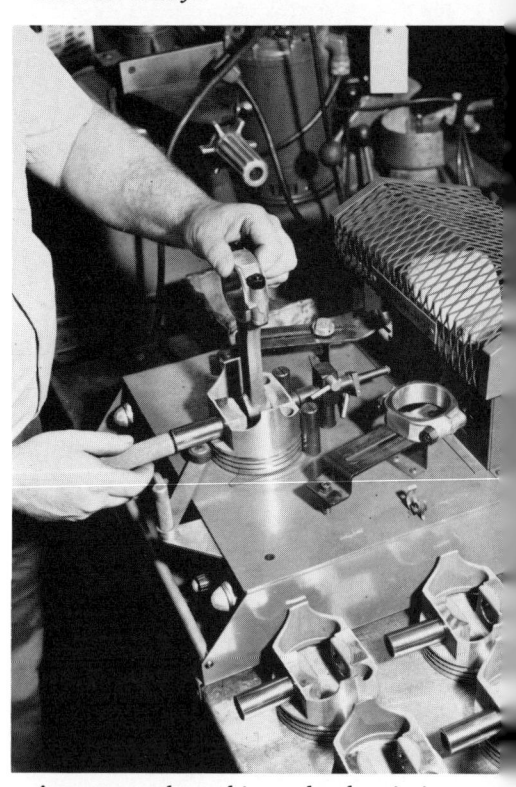

As soon as the rod is ready, the pin is pushed through to a stop on the opposite side. Fast work keeps the rod from chilling and hanging up the pin.

front, the wide side of each piston is toward the right, in the direction of rotation.

Pressed in pins are generally installed by the machine shop. The small

54 • PISTONS - RINGS - RODS

end of the rod is heated in a special oven. The piston is set on a fixture with a stop that will locate the pin and the pin itself is held ready by the fixture. When the rod is hot enough, it is taken out of the oven and the pin promptly pushed through. The trick is to move fast so that the pin does not hang up on the rod. The other trick is not to overheat the rod.

With full floating pins, the choice of locking clips depends on how the piston is machined. True Arc retaining clips are stamped, which leaves them with one rounded and one sharp outer edge. They are generally installed in pairs, with the rounded edges against each other and the sharp edges toward the groove walls, one out, the other in. To save time, preinstall a pair of clips on one side of each piston, oil up, push in the pin and install the other pair of locks the same way. Inertia forces do have an effect on the True Arcs and they need to be installed with the opening facing down or up.

Spiro Locks are threaded into the pistons, in pairs. Jim Cavallaro of Diamond Racing Products suggests making up a handy installation tool out of a screwdriver that is bent at right angles and sharpened to form a flat foot. Spiro Locks go in easy, once you get the hang of it and are very good about staying in. For removal, use a small screw driver or an awl.

Where the bevels at the ends of a wrist pin provide a wedging action and pin length is carefully controlled, a round wire lock is very effective. On the other hand a round wire lock used in conjunction with a square pin (the old fashioned way) is an invitation to problems, because the pin pounds out the round wire and gouges out the cylinder. Buttons at the ends of the pin are very positive in locking the pin.

With piston hung on the rod, check that the piston swings freely. You should not be able to feel play. Now you are ready to install the rings, on one piston at a time, to match the piston number and the ring set fitted for that bore. Begin with the expander for the oil ring. On a Speed-Pro ring it will have two color coded tangs which must face each other and not overlap. Next, the ring rails are installed so as to nest between the oil ring land and the expander. The rail gaps should face about 120 degrees apart.

The second and first compression rings are installed next. You can use a ring expander to slip them on, or spin them in. Major engine builders do it both ways. The main thing is not to overstress the the ring and bend it by mishandling. Use your thumbs on the ends of the rings to keep them from scoring the piston surface. The top rings are spaced 120 degrees apart, and you should avoid having the slots line up with the ones of the top oil ring rail.

Rings rotate all the time and the slots move. The chances of your finding the slots as installed are nil, unless the rings froze. Troubles begin when the rings don't rotate, and the science of exactly locating the slots should include the height of the moon at engine assembly time.

The popular Total Seal second ring is used by many engine builders, such as Ron Shaver, to control blowby. It has an L-shaped cast iron ring, the exact reverse of a Dykes ring. A steel rail is inserted into the hollow of the L to hug the bottom of the ring groove. Together they have the same cross section as a 1/16 cast iron ring but they work very differently. The top ring covers the top and back of the gap in the steel insert. By the same token the steel insert closes off the gap through the

The two compression rings go in with the dot up. Either spiral them in or use a ring expander.

The steel rails are now slipped in between the expander and the piston, one at the top and the other at the bottom.

Insert the oil ring expander into the bottom groove—it has two plastic marking tabs which butt against each other and should never overlap.

The expander pushes the steel rails outward. The telltale tabs are next to each other and the gaps in the rails are spaced away from each other by one third of the piston circumference.

PISTONS - RINGS - RODS • 55

We have the block and crank all cleaned and prepped and now we are going to install the pistons and rods. With pressed-in pins the machine shop will deliver the rods, pins, and pistons as an assembly.

top ring. That's why Total Seal calls it "Gapless."

The inside of the cast iron section acts as an expander for the rail. During initial break-in, the cast iron seats first and then the extra pressure helps seat the chrome rail. Gas pressure acts on the top and inside of the Gapless ring to increase the sealing action during compression and power. We assume

Push on the bearing shell with your thumbs to slip it in to the cap or the rod without scraping its back against a sharp edge at the parting line.

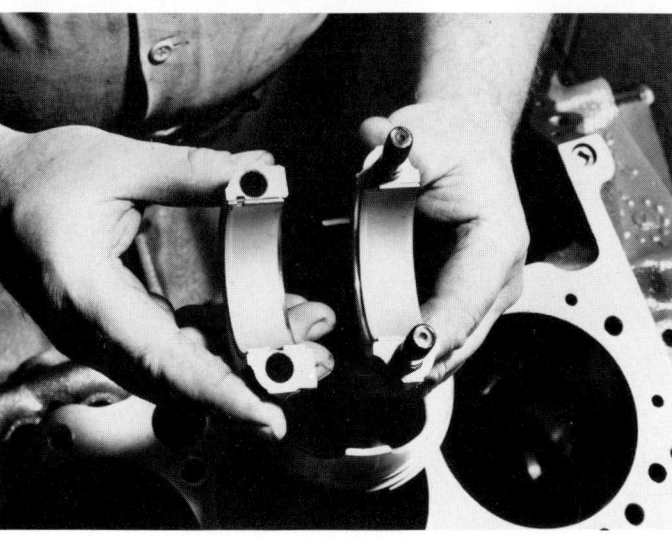

Top and bottom bearing tangs face to the same side. The cap and rod numbers must also match on the same side.

that there is some torsional action involved, because the Gapless ring seems to be good at sealing vacuum as well. The cast iron section is installed first, with the wide face up, and then the rail is set in. Keep the gaps at 90 degrees to each other.

BLOCK ASSEMBLY

Go through a final wipe down for the rod journals and cylinders. You should have a crank socket and a torque wrench on the nose of the crank. Get a torque reading of what it takes to turn over the crank and then get an extra torque reading for each piston and rod assembly as it is installed. This torque reading will increase with each extra piston in reasonably equal fashion.

There is horsepower in the block. CJ Batton and Jim Vincent are testing out their latest combination of head work and block machining.

56 • PISTONS - RINGS - RODS

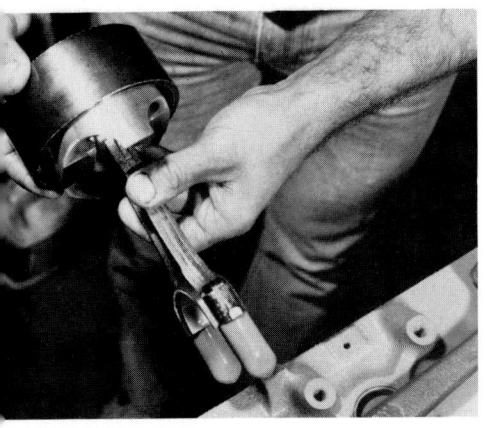

If you often work with one ring size, such as a 4.030 bore, a one-piece ring compressor is a time saver. Always use bolt protector boots to keep from nicking up the crank.

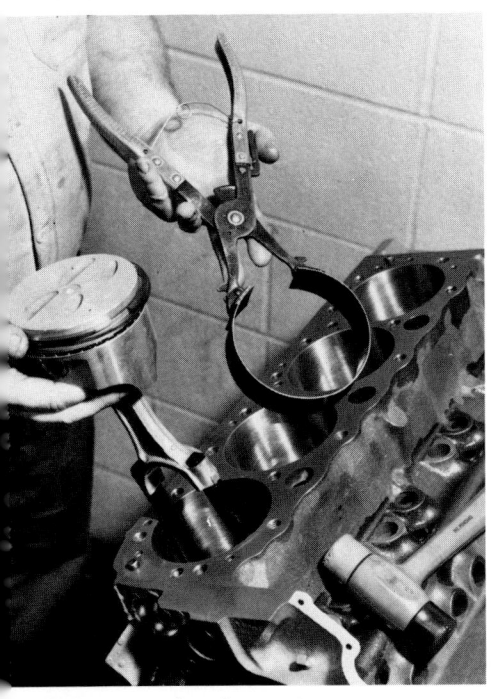

Either use a band type ring compressor with a squeeze handle or a wind up type with a key. The ring gaps should be 120 degrees apart. Rings rotate continuously while the engine is in operation, and never stay the way you put them in.

hold for another piston which meant a delay, plus balancing and detailing—but no shortcuts. Those who do take shortcuts, also eat a lot of engines.

Bearing shells are installed dry so the rod has a good grip on them. Use your thumbs to press on the bearing shell so it does not scrape against the edge of the cap or rod as you drop it in. Oil up the cylinder bore, the journal, the piston and rings, and install the ring compressor. If you have a band-type ring compressor tap against the sides with a plastic mallet to seat the compressor on the rings and recheck the tightness. The same thing goes for a squeeze type ring compressor.

Slip a pair of protector boots on the rod bolts, check one last time that the numbers and the tangs are facing out, and slide in the piston and rod assembly. Tap lightly on the ring compressor so it seats squarely on the block then drive down the piston and rod till the rod is on the journal. If a compressor is not seated, stop, backup and try again. Any hammering on the piston with a ring that has escaped the ring compressor, breaks parts. It helps to have the journal all the way down, so the cap slips on easier.

When you install the cap, the two bearing shell tangs will have to face the same way. If the cap does not slip on easy, don't tighten it till you check it. OK, the numbers line up, you tighten the nuts or the cap screws in a couple of steps, then turn the crank one full turn. If the torque increase is modest go to the next piston in the same journal. By the time you are done, all the pistons are in. Now retorque all of the rods one at a time. Turn over the engine and see that there is no increase in torque. Make a quick feeler gauge check to find the clearance between rods.

The trick is to work clean and not to embed particles into the bearings. If you are using mikes and dial bore gauges, your clearances are controlled well enough. Most engine builders prefer to use Plastigage as a final check either during final assembly or just before.

Gently tap the ring compressor to seat it against the top deck so that the rings cannot slip out, then send the piston through. If a ring skips out of the installation tool and expands, it will catch the deck and break if you hit the piston. After each piston is installed, turn the crank a full two turns with a torque wrench; check that the torque has not increased out of hand, and that nothing is binding up. "If it doesn't work, get a bigger hammer!"

Best way to avoid mistakes is to lay out all the parts on the bench with the rods and select fitted bearings as well as the ring sets. Install the rings a set at a time.

The piston and rod are driven home—guide the rod so it is in line with the crank and engages smoothly.

A sudden jump in torque tells you something is wrong—back up and find out why. If a bearing is too tight or a ring expander overlaps, or a ring broke or the piston didn't feel right going in, or a rod nut is not tightening right, it is up to you to stop, back up, and find out before it becomes back to bite you with a major problem.

We paid a visit to a professional engine machine shop just as the man was ready to put the last piston in the block. He happens to take the trouble to check the pin bores of each piston on a dial bore gauge and found one that was slightly off. The engine was put on

PISTONS - RINGS - RODS • 57

COMPLETING THE BLOCK

OIL PUMP—OIL PAN—WINDAGE TRAY

When a connecting rod or a crank throw collide with an oil droplet it is a case of instantly accelerating it to crank speed and this soaks up horsepower. Now multiply that droplet by the zillion to duplicate what is happening inside the crankcase and you are speaking of some serious horsepower loss. It's like putting eight quarts of oil through a blender. Switching from a stock pan to a specialty pan can gain 30 to 40 hp. Naturally there are differences between pans and we usually speak of a 5-15 hp bonus when you get a better one.

Adding a good windage tray and oil pan is like bolting on horsepower. The deeper the pan, the better it works; it allows the oil to fall away from the crank and rod assembly. The smart pans carry deflector baffles to help oil return to the sump. Some have screens with a directional action, others have louvered baffles. There are wide pans for stroker engines, pans with a kickout to let the oil sling out and drop into the pan. The reason for all this is to get the oil away from the crank and rods and to use up less power on just agitating the oil.

When you don't have enough road clearance but still need the extra oil pan volume, switch to a pan like the Moroso with a wing type sump. It is wide enough at the bottom of the sump to hold an extra quart or two, and can make a large difference in oil temperature and reserve.

Where large acceleration, braking, or cornering forces are involved, the pan must be designed with trap doors to control and channel the oil toward the sump pickup. It is a science handled by top people like Moroso, Hamburger, Canton, Aviad, and others.

WINDAGE TRAY

A windage tray serves as a shield and a shear plate—it takes some of the oil and air entrained with the crank, cuts it away and sends it into the oil pan. Moroso supplies a stud kit that holds the mains and supports the windage tray. During preassembly, adjust the tray so the rods clear.

Moroso's Tom Abbot explains that oil is entrained with the crank and rods. Looking at an engine from the front, oil from the crankcase swings from 3 o'clock down toward the pan and would now like to continue to turn with the crank. With a stock pan it is hard to prevent this. Now, let's add a pan with a kick out—an extension that projects well beyond the block. When the oil reaches this kick out, it smoothly redirected away from the crank and back down into the oil pan. You just saved some horsepower...

OIL PUMP

Extra oil pump volume overcomes all the leaks at the bearings and gains extra oil pressure. You also need pump pressure to feed the bearings and to cope with rpm and centrifugal force acting on the oil columns in the crank. The volume of oil going through a bearing cools it, a must for durability. However, it costs more power to drive the pump, which is relatively minor. It is also harder to control the extra oil being thrown around the crankcase and that costs sizable power.

Lee Bandrow from Lab Machine points out to another major disadvantage of a large pump: it can suck a crankcase dry. Some of the oil is going to hang up inside the engine and not return fast enough. Now, the big pump suddenly empties out the oil pan and you pump air instead of oil—good-by engine.

At idle or low rpm the engine responds favorably to a big pump, and gets up to full pressure earlier. You are happy with not seeing the low pressure, or the red eye as you are get off the freeway, but the problems happen at higher rpm. The answer is to combine a big pump with a big pan or switch to a dry sump system. After a big pump reaches the intended pressure, the pressure relief valve opens. Now the pump is doing work, but not accomplishing anything constructive because it just pushes extra oil out the relief. Size the pump to your engine combination.

People spend a lot of time drying up the top end to reduce oil flow. We don't agree. You need oil up there to cool the valve springs and help them live. Spend a little time on venting and oil drainage, and less oil will be trapped in the heads. Also, think about adding a little extra oil directed to the cam thrust surface and to the rear main.

You have several excellent pumps to pick from such as a Speed-Pro 224—4146 which is the stock replacement and the 224—4146 A that has the short gears but a higher spring pressure.

These Moroso main bearing studs support a windage tray. Height is adjustable. Cutting down on oil splash and windage saves horsepower.

The standard Chevrolet bypass adapter and oil filter mounting allows oil to get around a clogged oil filter. Some filters also have a bypass. Warm up the engine oil before you go racing and there will be less bypass.

58 • COMPLETING THE BLOCK

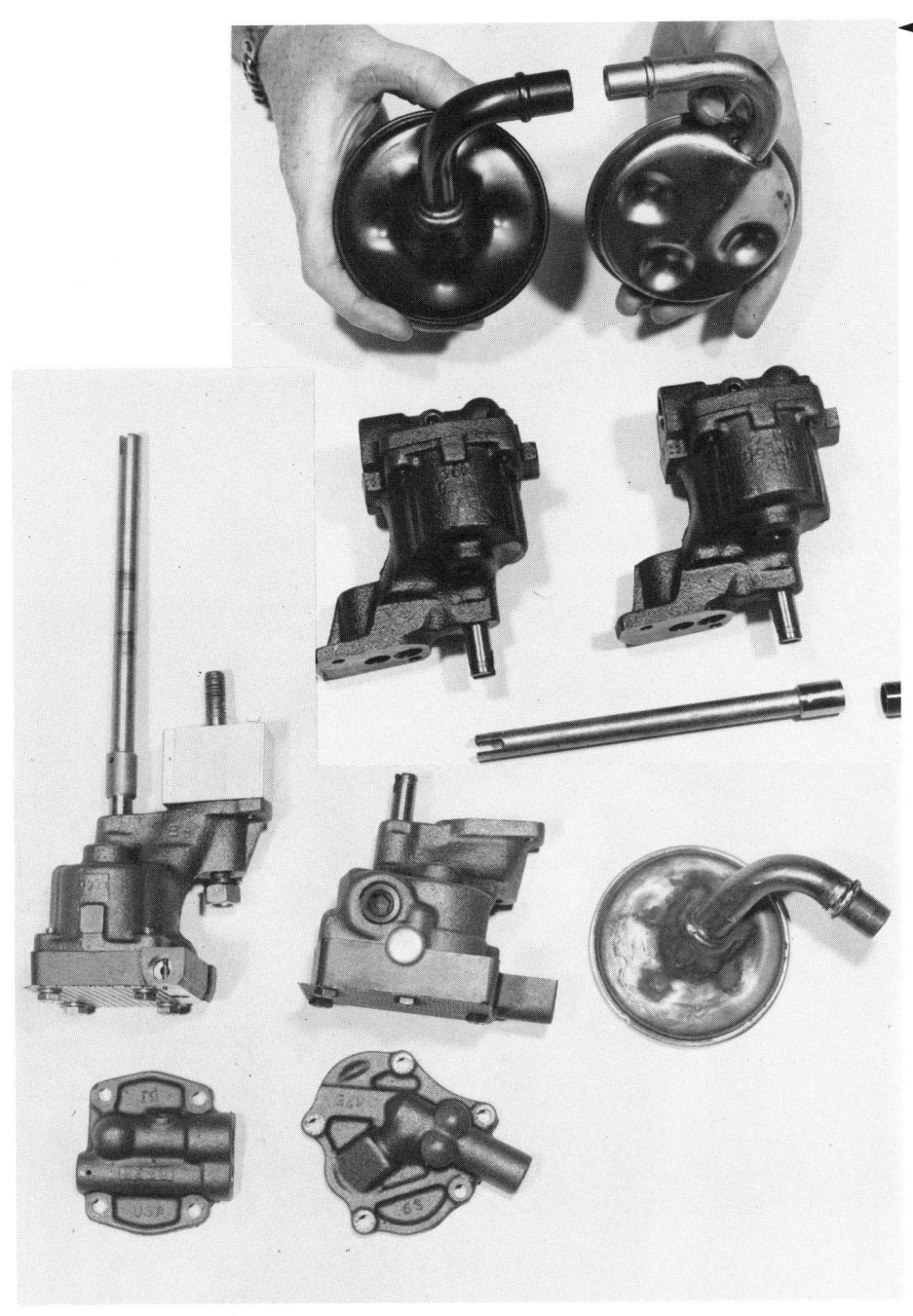

Use an oil pump that matches your engines requirements. With a small pan and a large pump, you stand a good chance of emptying out the pan, because some oil inevitably hangs up on top. Use Speed-Pros drive with a steel shaft and bushing. The oil pump pickup screen needs surface area.

This Moroso windage tray sends oil spraying from the bearings back to the pan, but won't let it spatter back up.

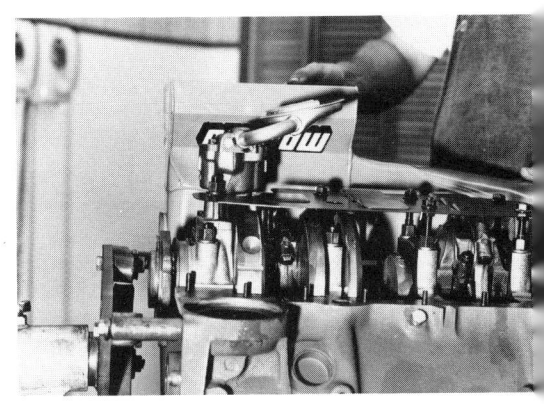

The pickup should clear the pan. Morosos stud kit at the pan rails holds the gasket in place.

Pumps with extended mountings and an integral pickup are available for different pan depths. Extended pickups should be braced separately to the pump cover.

Melling, Moroso, Weaver, Peterson, and Faria make excellent pumps. Melling specializes in cast iron pumps and gears. Steve Faria makes a special cover which allows both the drive and idler shaft to be fully supported by the housing at the top and by the cover at the bottom. He also adds a pressure regulator which is accessible through a screw in plug in the pan and is externally adjustable. The bypass is designed to return the oil directly to the pan. In addition to the stock gear length they build 1.125 and 1.400 gear length pumps with 12 tooth gears.

Speed-Pro packages a steel drive sleeve which connects the pump drive to the shaft and eliminates the nylon bushing. This 224 6146E shaft and steel sleeve are engine savers. The shaft must be slipped in before the pump is installed.

The pump is retained by a single bolt, which is best replaced by a stud, Moroso, ARP or similar. Also, the base of the pump is not matched to the main bearing cap. Coat the base of the pump with a little prussian blue, install it, bolt it down, then remove it. The imprint is what you match to on the main bearing cap. Do a little blueprinting to improve the oil delivery. Careful, there are no gaskets and you can't afford a leaker.

OIL PICKUP

Every pump needs a good pickup. Discard the old one, which is uncleanable and get a new one. The stock pickup is good, has the square inches of area, and the extra capacity helps keep it from plugging. The thing you find most of in a new engine is blue silicone chunks all over the screen. It goes there like drawn by a magnet. If you want to preserve oil delivery to the engine, use silicone sparingly so it will not get detached by the oil flow and circulated to the screen; also use a large enough pickup that will not clog prematurely.

The sump pickup must be lined up with the bottom of the pan. Measure the height of the pan, using a straight edge across it and another one down to

COMPLETING THE BLOCK • 59

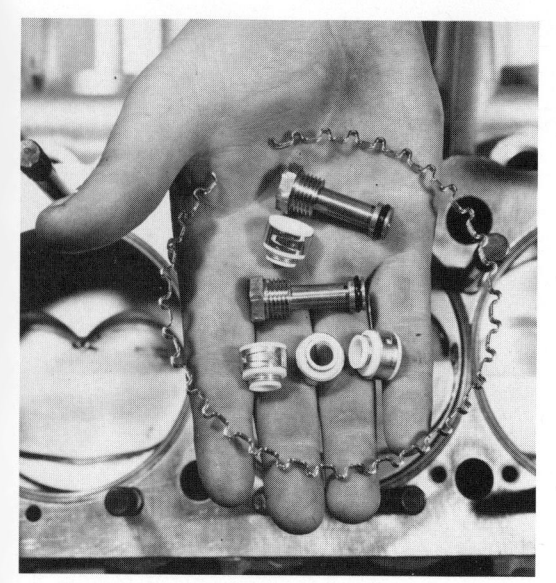

Oil going into the cylinder is controlled at the oil ring, the valve stem seal, and by limiting the top end oil. Bearing clearances also have a considerable effect and so do oil leaks into the intake manifold.

the block. Next, measure the height of the pickup, allow a quarter inch of clearance so oil has room to get in. The pickup tube can be held in place with green Loctite or Pro Loc IV, a little brazing, or a couple of spot welds. If you have a solid mounted engine or an off road application, brace the pickup. Speed-Pro carries an easy to use pick up installation tool that drives it in by the shoulder, so you don't need to hammer on the tube.

There are many after market pickups what range from Moroso to B and B or Hamburger and are tailored for deeper sump pans. Do your own inspection and check that they have enough screen area to suit you, and are the right length for the pan.

Most street or bracket racing applications use a wet sump system with all the oil carried in the pan. Jerry Weaver of Weaver Pumps in Carson City, NV explains that for oval track and fast bracket cars, going to a dry sump oiling system has specific benefits. The main one is that you can install a large capacity oil tank holding two or three times as much as a standard oil pan. Since oil is important to engine cooling, more capacity and longer cycle time is important. Most circle track engines need dry sumping, and Pro Stock engine builders use them to good horsepower advantage. Bob Rinaldi who builds Rob Moroso's engines sees a consistent 8-10 hp in using a four stage pump instead of a three stage pump.

A considerable amount of aeration and foaming takes place especially at higher rpm. Air is constantly being mixed into the air, yet air is not a lubricant-you want to get it out. With an oil pan there is limited volume, constantly recycled and the air doesn't get time to flow out. A dry sump pan and the big tank give the air bubbles time to work out of the oil. The oil picked up from the bottom of a tall sump tank is air free. Another advantage is you no longer drive the oil pump from the cam, which gets rid of cam thrust problems and spark scatter. Also with dry sumping you can switch to a steel rear main with no holes for pumps and oil galleries, hence more strength.

With dry sump pumps one or more scavenge stages are connected to the oil pan. The pan oil-pickups are normally aimed down toward the bottom and are protected by screens so that engine debris will not reach and wipe out the scavenge pump. One of the little problems involved is that debris not only wipes out the scavenge pump but can also stop the pump and pop the belt. A number of racers scavenge from a sealed off lifter valley as a way of keeping valve train pieces from getting down into the oil, and reducing the amount of oil getting into the pan.

The pump is generally mounted at the front of the engine on the driver's side and is driven by a timing belt. The high pressure side picks up from the bottom of the tank, connects to a filter, and then to the engine. The scavenge side delivers to the top of the tank, with the oil cascading down baffles and deaerating.

Scavenge pumps pull down a certain amount of volume per turn, and suck in not only oil but also air. This enables them to vent the crank case and also to pull down a slight vacuum, which is beneficial to oil control. Moroso's Bob Rinaldi points out that this allows less oil ring tension and can also save horsepower.

TIMING CASE COVER

The timing case cover forms the front sealing portions of the oil pan and must be installed first. Before installation, straighten it out at the bolt holes. Also drive out the front seal and replace it with a new one. It is easier to do it on a press, but an installation tool or sock-

This adapter allows the oil filter to be turned sideways, eliminating chassis clearance problems. It also carries the oil cooler connection lines.

Plastic pump drive collars disintegrate: use a Speed-Pro steel shaft and steel drive collar on any engine you build. The pickup must be pressed into the pump, aligned, and brazed in place or heliarced—also braced if possible. Some install it with an anaerobic sealer.

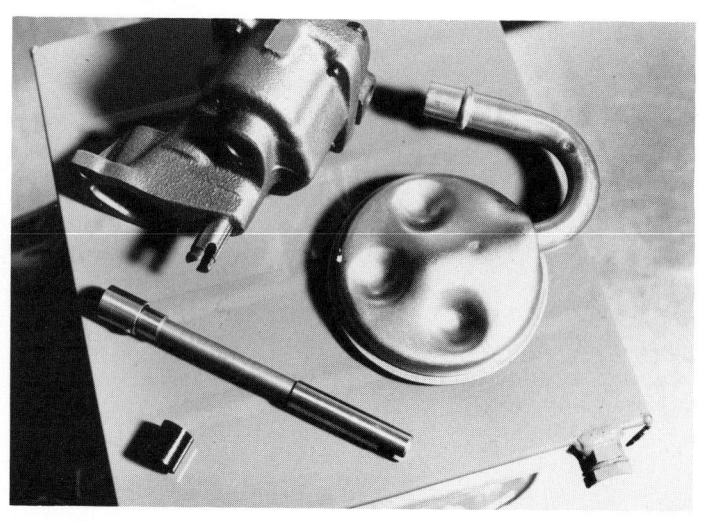

60 • COMPLETING THE BLOCK

ets and some tapping will achieve the same purpose. Apply sealer to both sides of the gasket and install.

If the original damper is grooved up, you have two choices. Either get a new damper, or install a damper saver kit. It consists of a small hard chromed sleeve which is driven onto the damper, after first applying a film of anaerobic sealer to the damper hub. Works like a charm. When there is no dowel pin, center the timing case cover from the damper hub before tightening the cover bolts. You will also want to do this before the pan is in place because that will throw off the seal location. If the seal is not centered it gets wiped out.

In a drag machine, the scavenge tank is mounted in the engine compartment to shorten the lines. In an oval track car, the tank goes into a protected area at the rear.

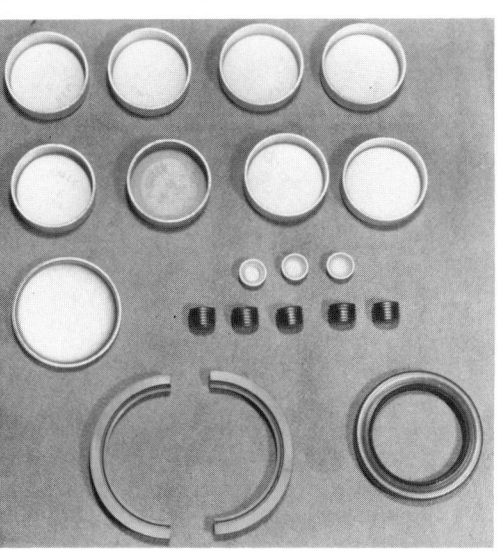

OIL PANS

Precheck the oil pan for clearance with the block and rods, especially with a stroker engine. Next, use the pan to set the pump pickup height. It should fit without interfering with the pick up. Your best bet is measure the height of the pick up and also measure the pan depth at the pick up. As a final check, try the pan on for size with the pickup in place and no gaskets. The pick up must be welded in place, without putting excess heat into the cover. You should also brace it. If it falls out, you just lost an engine.

Take time to check the clearance between the front and rear of the pan and the grooves for the front and rear rail gaskets. Often it takes a few taps to even out the spacing so that the gasket can do a better sealing job. On a pan that is being reused, straighten out the oil pan rails where bolts have dimpled them in. If possible convert from bolts to studs—they have a better grip and help locate the gasket.

Cement the pan rail gaskets to the block and set in the front and rear rail strips. Two different thickness front sealing strips are used on early and late Chevy blocks—measure and cement in the right one. A dab of silicone at each corner helps prevent oil seepage.

The pan side of the gasket does not require sealer, and later disassembly is easier without it. Run down the corner bolts or studs lightly to seat the pan front and rear, then follow up with the pan rail bolts. Don't try to bottom out all the bolts at once, just ease down and keep from bending the pan rails. Leave a little setting time and retighten. The oil level tube receives a dab of sealer and is tapped into the block. If it is loose, get a new one.

Once you have installed the pan, and made sure the drive key is in place, install the crank damper. No, don't hammer, thread in the installation tool, stud, thrust bearing and washer, plus a nut and draw it in place.

We are ready to turn the engine right side up and do the top end.

Speed-Pro replacement pump and screen have been welded at the correct angle after checking pan depth. High pressure pump is preferred. ➡

Use sealer at the freezeout plugs and drive them in with an installation tool or a socket. Avoid cupped plugs and replace them with threaded screw in plugs where possible.

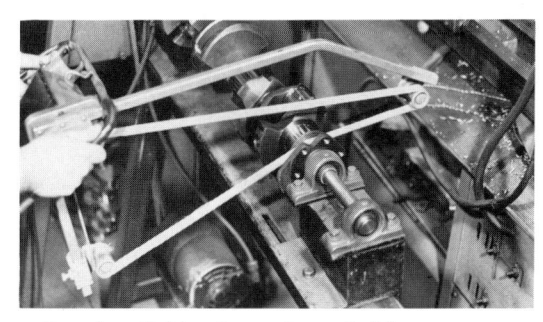

Crank polishing is not intended for removing material or straightening the geometry; it is only there to put on the surface finish and to make sure that the minute metal particles raised by grinding are smoothed in the direction of rotation.

This is the beginning of an ultimate billet 454 Chevy crank. It will allow more metal in the arms and counterweights than a stock forging.

Use a timing gear with an ample bevel to clear the front of the snout radius. If the gear hits, it forms a thin line that begins a crack. Your timing gear and damper keys must be a snug fit but not so tight that they cause a root crack in the keyway. Do not use the bigger hammer.

COMPLETING THE BLOCK • 61

HEAD PREP

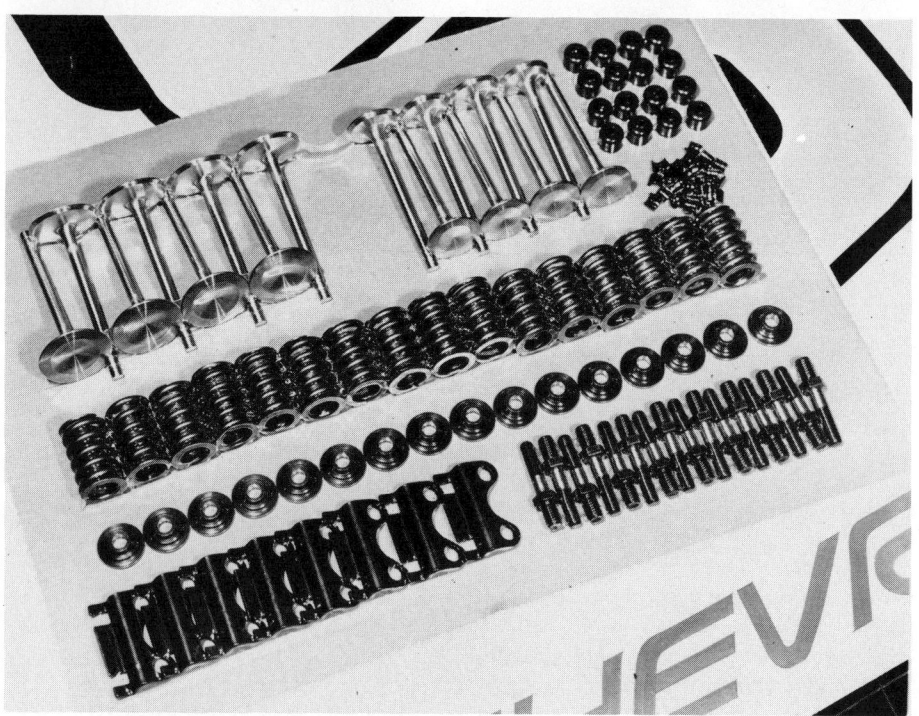

Stainless steel valves, hardened keepers, and valve stem seals, dual springs with flat wound dampers, spring retainers, pushrod guide plates and studs, there is enough there to finish up a set of 350 to 600 horsepower heads.

When you build an engine it pays to have the heads ready first, well before final assembly. Now, as the block is completed you can keep right on going. Your parts are all clean and laid out, so you save considerable time. Also, the inside of the engine stays cleaner. Once the heads are completed, oil them and keep them in a sealed plastic bag, till needed.

Here, we are going to take the heads apart, clean them, do all the measuring and machining, and them put them back together.

DISASSEMBLY

To remove the valve spring retainers, you will need a spring compressor. You can get a hand operated C-clamp type spring compressor for small money, or spend the extra dollars for an air operated one, which saves considerable time—much more than the difference in cost. The spring compressor arms need to be set to the width of the retainer, so that they don't slide off to one side. Clean and save the retainers and keepers, and scrap the valve seals.

Varnish, particularly in an older engine, will cement the valve spring retainers and keepers, which places an unnecessary strain on the valve spring compressor when you try to break them loose. The easy way is to first jar them loose—use a one inch socket over the valve spring retainer and hit it with a hammer. If the stuff is really frozen, use a supporting block or socket under the valve and then jar the retainer with a tap on the other socket.

The valves should now be free to to slide out, but they usually won't. If you start banging on the valves with a slide hammer, they bend and the guide is damaged. Instead, simply fix the keeper groove area. When keepers and valve have been pounding against each other for a long time, slight burrs or raised lips are formed. A mild clean up with a small file, or a strip of emery cloth drawn back and forth across the keeper area allows the valve to slip right out without scoring up guide.

A head needs a good cleanup before inspection, but there are some things you should note. Were the head gaskets sealing? Do you see burn or leakage marks between combustion cham-

Spring shims are used to achieve the correct individual installed height at each spring. Keep the shim stacks and the springs in sequence to match each location.

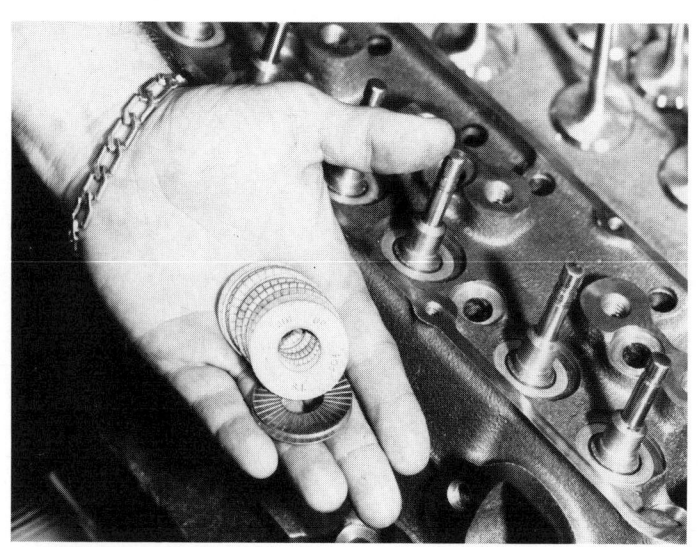

A telescoping snap gauge is an accurate way to measure available height between the retainer and the spring seat.

62 • HEAD PREP

The top of the valve guides is cut to accept valve stem seals on Bill Mitchell's Tobin Arp machining center. This head is a Dart II.

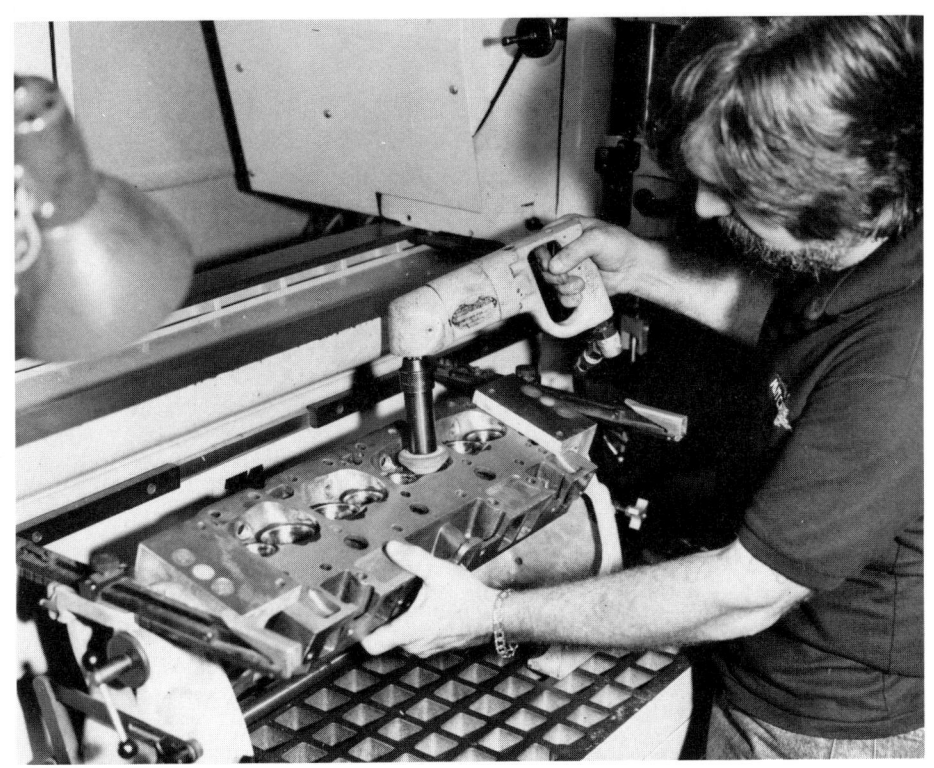

Bill Mitchell cuts the seats to be just a shade smaller than the valve. You want to maximize seat life and still gain top flow.

This tool cuts the valve spring seats to accept a larger diameter spring. Note that there is now less room for the center bolt—you may need a special cap screw or a stud with a 12 point nut.

bers or around the head gasket area? Check out the combustion chambers for signs of water leakage, and eveness of burn. Did the chambers show signs of excess oiling? Those little warning signs will help your planning and may tell you the head needs decking and probably guide work.

The cleanup, in a degreasing tank and bead blaster will help reveal any cracks. Do this crack checking before you plow in a lot of time and money into a set of heads. It is cheap insurance. On a cast iron head, a Magnaflux check is a more powerful crack checking tool than visual inspection. Brushing out the guides eliminates varnish build-ups and allows a clearer check of guide and valve wear.

VALVE AND GUIDE INSPECTION

Valves are cleaned on a wire wheel. First step is to see if the valve can be reground. Check the seat for wear—how much cutting will it take to restore it? If the valve is burned or cracked, it must be discarded. Also, the valves need enough thickness above the seat—margin—to keep them from burning. If the top of the valve forms a sharp edge with the seat, the valve is useless. You want that margin. In a strong engine, use stainless valves with hardened tips on intakes and exhausts, and seriously consider at least exhaust seat inserts.

Compare the wear at the portion of the valve stem that runs in the guide with the unworn top part of the stem. High wear means it is time to replace both the valves and the guides. The convenient standard check is to insert the valve in the guide and shake the valve. Guide fit will range from 0.0012 to 0.002 inch. Since the valve extends above the guide it amplifies the motion and you'll feel more looseness.

CONVERTING TO UNLEADED

Picture an absolutely clean valve and an equally clean seat. As the valve bangs down, small miniature welds form at different spots between the valve and the seat. Next time the valve lifts, those tiny welds break loose. Now the valve comes down again and the process repeats. As this repeats, the valve seat tends to sink. In effect, the valve augers into the seat which does little for flow or sealing.

Valve specialist Alan Reed from Flaming River Industries in Cleveland explains that when you switch from leaded regular fuel to unleaded, you get ripped off not only on octane value but also on the ability of the fuel to leave enough deposits between the valve and the seat and this valve seat recession becomes a serious problem. When you see the seat lowered by 0.060 to 0.100 inch in the head, you won't have any doubts as to what happened. Since regular fuel now unleaded, or given a minimal lead content, you should really convert to unleaded. It's *easy*—install exhaust seats and stainless steel intake and exhaust valves. Later heads do have induction hardened seats and suffer less from valve seat recession but by the time you get them and cut for new seats or larger valves, mighty little is left of the hardened seat. Kits for converting to unleaded fuel are available.

Milt Olson of Speed-Pro suggests that while you are at it, you should go for the extra power by switching from 1.94 intakes and 1.5 exhausts to the larger power making 2.02 intakes and 1.6 exhausts. Alan Reed adds that the larger wider inserts will not move around as much as thin ones thanks to the extra mass and support area. The final say so belongs to the casting. Some castings have very little supporting material and a crack can occur during machining through no fault of the shop doing the work. All aluminum heads carry valve seat inserts of varying size and durability. Thin ones do not last as long. Also sharp corners on the OD of the seat and in the machined pocket in the head should be avoided at all costs. For durability the valves you pick should be hard chromed along the stem. They generally receive a flash coating of chrome that is very thin but improves durability by considerable.

The cylinder head you pick will determine the engine's power level. Light 305 heads, while readily available, are not noted for durability. Yet, they will run OK if you switch to a large

All valves are seated to the same height to gain even cc in the chambers. Cc all chambers before cutting the head.

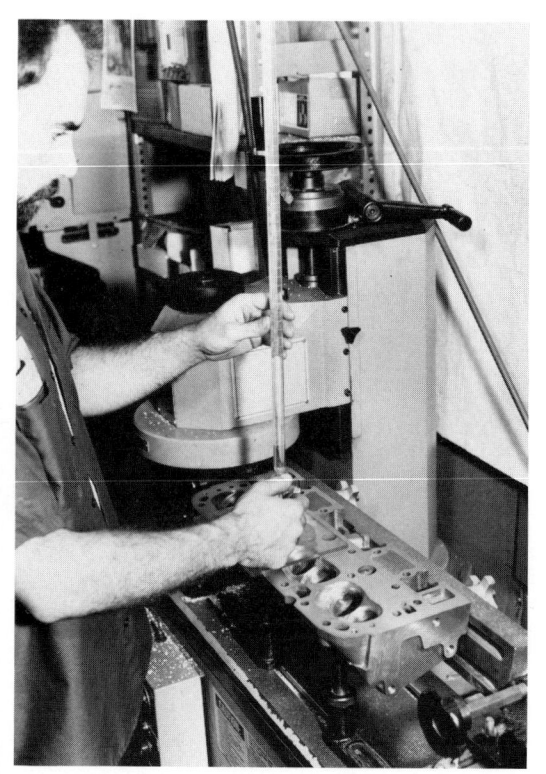

An accurate pilot slips into the guide. Valve seat accuracy depends on being centered on the guide.

radiator and do not allow them to overheat.

Ralph Thorne of South De Kalb Machine Shop in Decatur, GA showed us a neat porting procedure and head combination in which a 305 head provides a small cc conversion on a 350 Chevy. You can run more compression, use small valves, and pick up on economy.

Bob Lyon in Orlando, FL does a 305 head with extra large 2.02 intakes instead of a stock 1.78 to create a swirl head. It provides not only exceptionally smooth driving torque but also good economy. Your biggest gain is from being able to use the compression.

Older 350 heads come in innumerable versions and the best ones are now over 25 years old and crack prone. R.H.S. in Memphis, TN tells us that one third of the 461 cores they begin with are scrapped after the first disassembly and cleaning.

The Dart II is a cast iron head designed to replace the 461 with better ports and chambers. It comes with exhaust inserts and has better breathing out of the box than most reworked used heads plus extra metal in the right places.

Any used head should be checked for straightness and cracks before parting with the cash. Replacement heads range from Bowtie to Brodix and from cast iron Dart II's to aluminum ones. Cast iron is good at holding heat, but weighs more; aluminum is repairable, lighter, and easier to port, but is not as strong as iron or as good at heat retention.

VALVE STEM SEALS

The most elemental valve stem seal is of the umbrella type. It works by shedding oil away from the guide. It is easy to install, imposes no drag and is

During engine preassembly check for pushrod to head clearance and touch up the head if necessary. It is easy to break through—don't overcut.

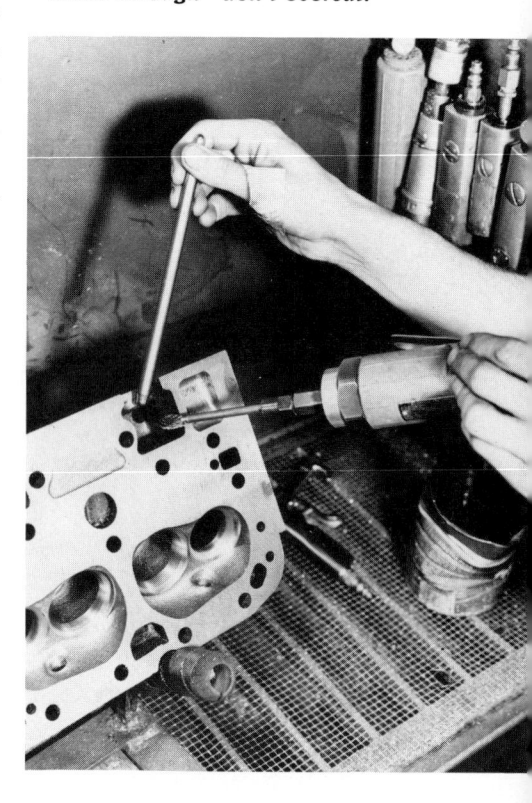

64 • HEAD PREP

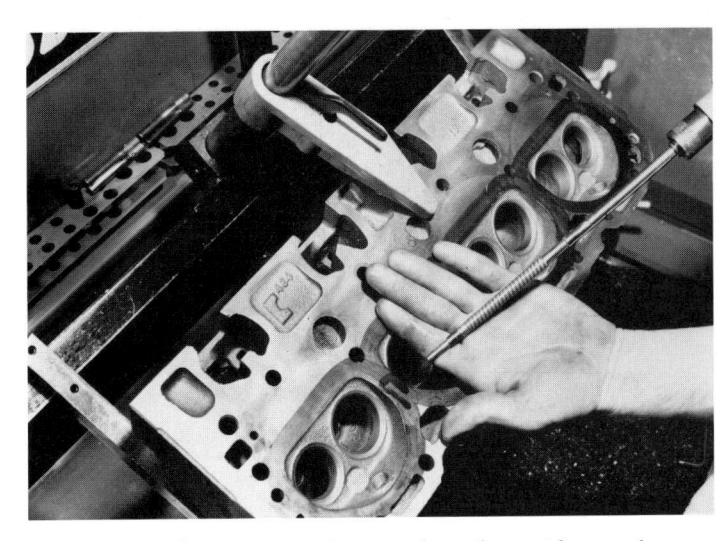

This Sunnen hone accurately sizes the valve guide to valve stem clearance. The smaller the guide to valve clearance, the longer the valve job lasts, as long as the valve can move freely.

R.H.S. designed this reduction drive with a chuck to grab the valve head—it allows polishing the stem for improved guide life. The flash coat of hard chrome is ultrathin!!

generally used with rubber O-rings at the retainer, below the keepers. One limit on the umbrella seal is that it cannot be used with inner valve springs. Also, it is not the world's most durable seal. Heat hardens the rubber and it eventually cracks and breaks.

Teflon seals fit on a machined surface at the end of the guide, hug the valve stem and wipe enough oil to just leave a metered film. It is definitely the superior seal, with a good installation, providing the guides clearances are within limits. Loose guide clearances knock out the stem seal.

Speed Pro engineer Cal DeBruin points out that an all Teflon seal is quite rigid. To seal it against the valve guide, use a dab of RTV. When there is enough room, use a rubber seal with a Teflon insert such as the ST2000, which allows the insert to self center on the valve stem.

There is a definite installation sequence for valve stem seals. First a tool centers on the guide bore and machines the top of the guide. This makes the OD of the guide concentric with the valve stem and provides a snug push fit for the seals. The valve stem seal adds a little height to the top of the guide. Clearance is needed at max valve lift to keep the valve spring retainer from contacting and destructing the seal. The distance between the retainer and the top of an installed seal should be larger than valve lift by at least 0.060. Bent pushrods—oil leaks at the guide—banged up seals are all reminders to check the retainer to guide clearance.

VALVE AND SEAT WORK

Valve seats serve as the exit for the intake port or the entrance for the exhaust port and have considerable effect on port efficiency and flow. An average valve job will provide you with a good seal but the flow work will depend on the individual shop doing the job. At the bottom end of the scale is a single 45 degree angle cut. The better way is a lead in or topping angle of 30 degrees and a 70 degree exit angle. There is no magic in a 3-angle valve seat, just extra work. Adding top and bottom angles narrows the seat and improves the flow. The seat should have an OD just slightly smaller than that of the valve. This way when the valve works against the seat for a while it will not start to sink into it. If the shop doing the work learned from their flow bench testing, the valve work improves and the engine makes noticeably more power. For instance, all ports and seats do not necessarily like the same angles.

Machine the valve spring pockets to the spring diameter. When checking dual valve spring pressure, use a washer at the inner spring to duplicate the spacing action of the spring retainers inner shoulder.

Umbrella style valve stem seals shed the oil so it will not drown the guides. They are used in conjunction with O-rings at the valve spring retainers, under the keepers.

HEAD PREP • 65

This tool cuts the valve spring seats to accept a larger diameter spring. Note that there is less room for the center bolt—you may need a special cap screw or a stud with a 12 point nut.

In times of old, valves were lapped against the seat as the final valve job step. A thin coat of lapping compound and a twirl stick or a hand-cranked lapping tool was used for the purpose. Both are still in my tool box—museum pieces with fond memories. With today's precision valve and seat cutting equipment, lapping the valves simply destroys the fine seat finishes.

However, valve lapping compound is very handy for marking seats. Apply some red or blue Dykem and the lapping compound marks the exact seat location with a thin gray line. All you need is a light touch. Now you can port or blend to near this line. Just allow enough room for the bottom cut of a 3-angle valve job.

VALVE SPRING HEIGHT

Spring pressure helps close the valve, and returns all the valve train parts. Weak econo springs cause valve float, cost power and destroy engines. Install the springs which belong with the cam you are running and set them at the correct installed height and pressure.

When the valve and the valve seats are cut, the valve sinks a little deeper and the stem protrudes further from the guide. This moves the valve spring retainer further from the spring seat. Now less spring tension is available because the installed spring height has increased. You compensate by adding shims under the spring.

The job can be done by just measuring heights, but then you operate in the dark—close but no banana. The better way is to use a spring checker which simultaneously measures height and pressure. You know from your cam card the spring pressures required, seated and open. Now you check on the spring checker the height at which this seated spring pressure occurs with the springs you have on hand. This is the installed height you work to.

A spring has a free height—the height at which it stands with no load. As you begin to compress the spring, for every new height there is a specific pressure. You would like to believe that all springs in a given batch are the same and that if you check one spring all the other ones would be just like it, but not in the real world. You can assemble a street engine that way and never turn fast enough to know the difference. To get serious and move up in rpm with a stronger cam and bigger valves, valve spring pressures really count.

Install the valve and spring retainer without the spring and check the height from the retainer to the spring seat. If that height is greater than the installed height you need, add spring shims to make up the difference. Spring shims must be fitted to each valve spring individually, keep the stacks separate so they can be installed in the correct locations.

A number of different spring diameters are available ranging from 1.25 to 1.440, 1.550 and 1.6 inches. The smallest ones are for stock street use and the larger ones are used for stouter machinery with more radical cams, roller or flat tappet. When you go to a larger spring, remachine the valve spring seats, using a tool available from cam manufacturers for the purpose.

Many heads lack material in the spring pocket area—you machine the new seats and cracks develop either right now or shortly thereafter, right into the water jacket or into the port. Depending on the spring combination,

The valve stem and the valve seats are oiled to keep them from rusting. Oil also helps start up by adding to the initial seal.

66 • HEAD PREP

the available installed height may be less than what the spring pressure calls for. If you are in doubt, don't machine the spring pocket and create water wells. Instead, switch to a longer valve stem or use a deeper valve spring retainer.

Coil bind destroys valve springs—when you open a valve far enough so that the coils of the spring bottom against each other, you destroy in quick succession the spring, the cam lobe, and all the parts in between. To prevent all this takes one simple check. Put the spring in the spring checker and see at what height the spring stacks solid or coil binds. You will know that this happened when the spring pressure suddenly shoots up, without the spring moving any further. It is this excess pressure that bends pushrods and destroys pieces. Picture the installed height as the starting point. When we now open the valve to its maximum lift point, the remaining spring height must be taller than the coil bind point by at least 0.050 inch.

You can check installed spring height with a scale—very inaccurate. The tang of a dial caliper can be used to measure height but it invariably sits at an angle. Most people rely on a telescoping snap gauge—it is set and locked in place, then read with a caliper.

B-H-J makes an excellent gauge that gives a direct reading with a dial indicator and a caliper like arrangement. Two slotted hoops at the ends of the caliper clear the guide and the stem to engage the spring seat and the valve spring retainer. Also available is a direct reading gauge which fits over the valve stem and screws out to snug up against the spring retainer. Read the marks on the gauge.

FINAL ASSEMBLY

All of the valve seats have been cut and the head has been cc'd with the valves you are going to use. This is very important because the thicker replacement stainless steel valves can occupy 3 cc's more than the thin stock valve adding a lot of compression you never expected. After the heads are cc'd, they can be milled for the desired compression. Check the Compression Section.

Don't overlook the valve stems in getting the heads ready. Clean up any burrs or high spots in the keeper groove areas and smooth out the stems—some coats of hard chrome may need polishing. Scooter Brothers showed us a neat reduction drive fitted with a chuck, which R.H.S. uses to grip a valve by the head and spin it while the stem is being lightly polished to eliminate all burrs. The system is used to clean up all their valve stems, new or old so as to ensure longer guide and valve seal life. One word of caution: the hard chrome coat on the valve stem is quite thin....

After a final washdown, with the guides brushed out and the valves cleaned, apply a coat of oil to the seats and at the valve heads. This prevents rust and insures a good initial seal at the engine start up. You also need oil or GM EOS assembly lube at the valve stems.

To install valve stem seals use the plastic cup supplied with the pack. It slips over the valve stem tip to cover the sharp edges at the keeper grooves. Press the seal onto the guide shoulder with a socket or a tool. If a valve stem seal is scored during installation, put it directly in the trash can and replace with a new one.

High intake manifold vacuum tends to suck oil into the guides. LAB Machine's Lee Bandrow points out that with a high intake manifold vacuum you can combine valve stem seals with the O-rings to cut down on oiling. Where there is less intake manifold vacuum, using the O-rings can dry out the guides too much. Some performance valves are not cut for O-ring seals.

Now use your valve spring compressor, install each spring with the corresponding shims, slip in the keepers and release the valve spring compressor. A rap with a raw hide or plastic mallet on the valve stem tip seats the retainers. Any keepers which were not properly installed will pop loose on the assembly bench instead of in the engine.

ROCKER STUDS

Most standard heads come through with pressed in studs, more or less adequate for stock springs. With any sort of performance, neither is worth using. Your best bet is to either buy heads with studs or convert to studs. They are available in 3/8 or 7/16 and the latter are preferred for any serious roller tipped rocker assembly. The studs are installed with the pushrod guide plates under them. Later you will want to center the guide plates to align the rocker arms against the valve stems, so the studs are fully tightened later.

Pack up time—oil up the heads and valves, put them in a plastic bag and seal up so they stay clean. When the block is built, the rest of the engine will fly together.

A quick tap on the valve stem, using a plastic hammer, will seat the keepers and reveals any that were ready to pop out.

Air powered spring compressor saves a lot of time. Make sure the keepers are fully seated and at even heights. If you have the least doubt, check them again. A valve with no spring control can take out a complete engine.

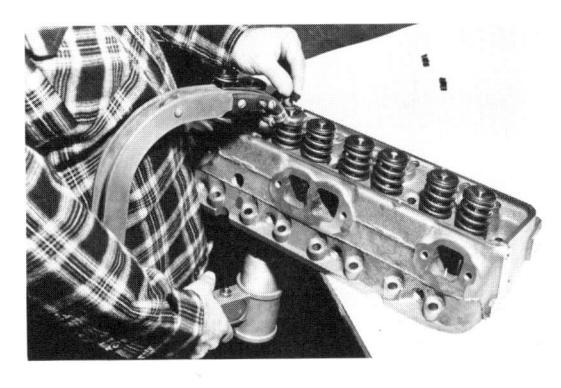

Set the spring compressor arms to fit the retainer.

HEAD PREP • 67

TOP END ASSEMBLY

INSTALLING THE HEADS

With the block completed we are now ready to unwrap the assembled heads and install them. We'll assume that the threads in the block are clean. Your head bolts should be cleaned, wire brushed, and laid out on the bench next to the engine stand. If the threads do not look sharp, they are worn—replace the bolts. Next to them are the hardened washers which you should be using to protect the cylinder head and

The head is located by a pair of replaceable dowels. You can remove them with a puller, a pair of vise grips, or with a claw hammer wedge.

spread the load. With hardened washers you will need longer bolts. Bolts or studs should engage the block by at least 1.5 times the bolt diameter. Physically check the head gasket against the block and against the head to see that they do not overhang into the bores, and that coolant holes correspond. Also, check the dowels in the block. You can purchase and install new ones.

When you first install the bolts, the shoulder sticks up, giving you an indication of the engaged thread length; make sure it is adequate (0.75 inch). Bolts are always tightened in the sequence indicated in the shop manual and torqued to 20 ft.-lb. initially to settle in the head. Keep retorquing till you reach the value indicated in the shop manual or by the bolt manufacturer. Increase the bolt torque to spec in 10–15 ft.-lb. increments.

Sealer at the head bolt threads will do double duty as both a sealer and a lubricant. You need oil or a moly grease at both sides of the hardened washers. Lock in studs with an anaerobic locking and sealing gel such as Fel-Pro's Tight which does double duty by locking and sealing the stud.

Even though a gasket may not need retorquing in a street application, it is good practice, after initial engine break-in, to let the engine cool and to retorque the head bolts, mandatory in a race engine.

LIFTERS

With the heads in place we are going to install the valve train parts, beginning with the lifters. Always inspect the bottom of the lifter for wear marks. Speed-Pro suggests placing two lifters bottom to bottom: if they rock, the surfaces are convex and the lifters are good. If the lifters do not rock, the surfaces are flat and the old lifters are history. With any cam change use new lifters. If you plan to reuse the cam and the lifters, keep the lifters in sequence using a block of wood with two rows of holes and a "front" marking. Each lobe is bedded to its corresponding lifter and this allows you to keep them together. With a new cam the new lifters go in any sequence.

The bottoms of the lifters call for rubbing in a light coat of moly based–cam assembly lube or Sealed-Power's LL-5. The sides of the lifter are oiled with GM EOS assembly lube. Lifters must be free to move up and down. It's easier to bleed out all the air from hydraulic lifters by prefilling them. Seaport Automotive uses an oil can with a small hose tip to "inject" the oil. Oil inlet faces up to let the air bubbles escape.

Hydraulic lifters vary in leak down rate. Standard ones are designed to simply maintain a zero lash and to leak down fast enough to avoid pump up. When you use higher spring rates, the leak down rate increases, which assists the anti-pump up action. Also available are high leak down rate lifters which provide a shorter timing at low rpm. As rpm goes up, the leak down time is reduced and you get more valve lift—more area under a lift/time curve

Heads are tightened from the center on out in a series of increasing circles. Begin at 20 ft.-lb. and work your way up so that the head beds down evenly.

and effectively longer timing. It sounds good in print and it offers a limited advantage in some classes, but you won't have any trouble becoming aware of the accompanying noise level. When parts are noisy and clattering, stress level increases. In effect, the extra leak down loses part of the cam's ramp action and we are not at all convinced that this is good, where the engine is intended for durability.

68 • TOP END ASSEMBLY

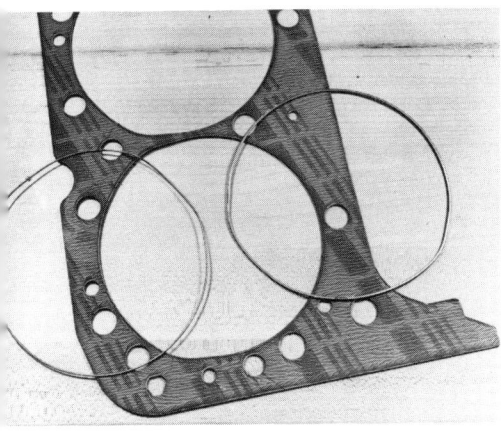

Fel-Pro gasket with a steel wire sealing ring is the best for a cast iron head. For aluminum heads use the Fel-Pro gaskets with the copper sealing ring—the wider and softer copper wire is more compatible with aluminum.

With roller lifters, always inspect the surface finish of the roller and also check the play of the roller at the pin. A roller does not live long if it cannot stay square with the cam, so check the roller alignment mechanism.

PUSHRODS

Make it a point to check all of the pushrods for flatness by rolling them on a flat surface. If you see the middle of the pushrod moving up and down, replace it. Pushrods seldom "just bend". Any time you notice a bent pushrod, look for the problem source. It can be a coil bound spring, a retainer hitting a valve seal, or a valve getting kissed by the piston. Always blow air through the pushrod to see that it is clear and can flow oil to the rocker arm.

Set a can of moly lube into the tappet valley and use it for dip coating the pushrod tips. Insert the pushrod so it fits through the guide plate or through the guide hole in the head and engages the socket in the lifter. Polish the guide plates

The block surface has been remachined and all of the tapped holes chamfered. The teflon coated Fel-Pro head gaskets are installed dry. Check that all the appropriate holes line up. The gasket provides coolant metering.

so the pushrods don't scuff, and apply moly to both. Surplus moly lube will land on the oil filter and plug it, so the filter should be replaced after break-in.

ROCKER ARMS

If you are using bath tub type rockers, check both the valve stem tip and the pushrod end for wear. Also, check for metal pickup in the rocker and at the ball. The ball and the rocker bed into each other and should remain mated. The easy way to clean them as a group is to string them all on a coat hanger or a piece of welding rod. With roller tip rockers check the rollers and the trunion bearings for wear.

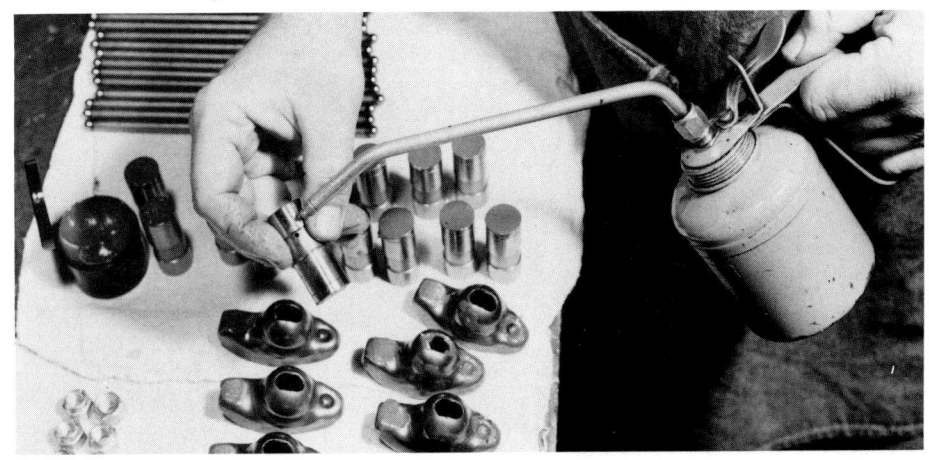

You can pre-fill the lifters and bleed out the air ahead of time to prevent start-up clatter. During prelube make sure all the pushrods feed oil.

VALVE LASH ADJUSTMENT

Adjusting valve lash on a Chevy engine is easier than on any other because the rocker arms can be moved up or down the stud. Stud nuts with

Coat both ends of the push rod with moly lube before installation.

Lay out your valve train parts on a table next to the engine being built.

locks are much preferred because they are better at holding adjustment and also accept stud girdles.

To adjust valve lash it isn't enough to have the valve closed: the lifter must also be down on the base circle or the no lift portion of the cam. With a very mild cam turn the engine till you have opened and closed the exhaust, then opened and closed the intake and come up to the firing position. Now, both valves are seated—more or less.

Competition Cams explains that the trick in adjusting the valve lash is to make sure the lifter is riding against the heel of the cam-just opposite the max lift point. This way you are not taking chances that the lifter is riding on the early part of the ramp and taking up part of the clearance. That is exactly half a cam turn or one crank turn from the max lift point. To get there:

ADJUST THE EXHAUST WHEN THE INTAKE IS CLOSING.

ADJUST THE INTAKE WHEN THE EXHAUST IS OPENING.

The system is very accurate. You can save time by adjusting all the exhausts in the firing order, turning the engine exactly 1/4 turn between adjustments, and then repeating for the intakes. The Chevy firing order is 1—8—4—3—6—5—7—2.

TOP END ASSEMBLY • 69

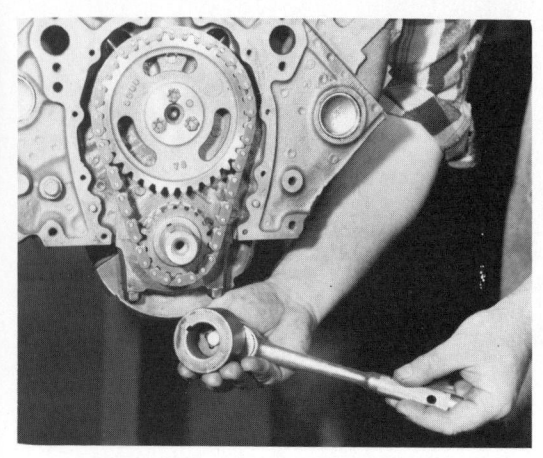

To turn the engine while adjusting the valves here is a handy Moroso socket to fit the crank nose.

Sight along the side of each rocker, with the valve up and see that it clears the spring retainer. Also, check that at max lift, the pushrod clears the rocker.

When the engine is done, protect it with duct tape and a plastic bag, till ready to install in the car. Murphy's Law: Anything left open will have something fall into it.

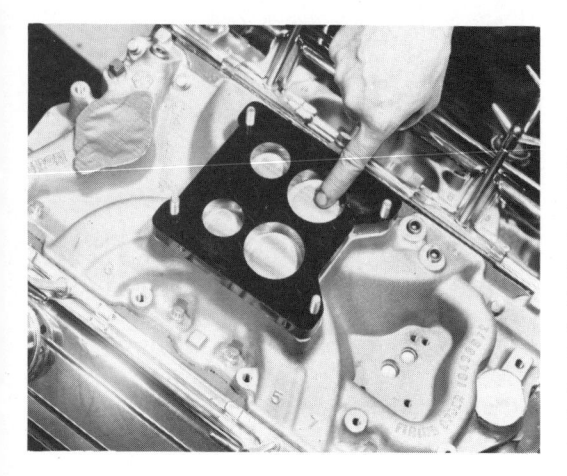

The lifters receive a coat of moly along the base as well as a coat of oil at the sides. With a new cam use new lifters. If you are retaining the old lifters, keep them in the original position.

The Chevrolet Shop Manual offers a different time saver for hydraulic lifter cams only: bring the number 1 cylinder to TDC (and firing) then adjust

Exhaust Valves 1—3—4—8
Intake Valves 1—2—5—7

Turn the engine one more turn and set

Exhaust Valves 2—5—6—7
Intake Valves 3—4—6—8

HYDRAULIC LIFTERS

With a hydraulic lifter, adjust to a zero clearance, then add one turn of the adjusting nut. This brings the plunger in the lifter down away from the retaining clip and leaves enough room for the plunger to move either up or down to maintain clearance and compensate for future wear. The adjustment begins by taking out all of the play, still leaving the pushrod free to turn. Going one more thread pitch gives the correct adjustment.

MECHANICAL LIFTERS

Your cam card will have the specs for setting valve lash. Use a feeler gauge of the correct size between the rocker and the valve stem and adjust the rocker till the feeler gauge moves with a very slight drag. A heavy drag reduces clearance and leaves you with no control. Valve lash can be set cold or hot. Hot settings at operating temperature sound great but are not necessarily more accurate. As the engine warms up or cools down, different parts of it will retain more heat than others, and the expansion of the block, head, pushrods, and valves vary all over the map. With the engine cold the parts are all at the same temperature. Cold and hot valve lash are not necessarily the same.

MORE TIPS

The tip of the rocker arm and the valve stem constantly rub against each other and wear, which results in dips and hollows. When you insert a straight feeler gauge it bridges the gaps and you read less clearance than there really is. Now that valve sounds noisy. Reducing the hot lash clearance at that location gets rid of the noise. You can test for a noisy valve by inserting a feeler gauge. If the noise goes away, try reducing lash.

Tight valve lash can burn valves, and will cause a distinct miss. If you get rid of the miss by loosening up the lash without getting into noise, you cured the problem.

As valves run and pound against their seat they tend to sink in, which reduces the valve lash. Also, if a valve gets sufficiently hot, as in the case of preignition, it will tend to tulip and sink in. The seat is holding it, the spring is pulling it, and with enough heat the head of the valve tulips. Valve lash loss is a major warning sign of heat—time to back off on the timing or to richen up. Conversely, increasing valve lash is a wear indicator. Check valve for valve lift change and wear at rollers in the lifter or in the rocker, or a pushrod trying to auger through a rocker arm. A

70 • TOP END ASSEMBLY

The rocker receives a coat of moly lube in three places: at the tip, at the ball socket, and at the pushrod end.

change of more than 0.002—0.003 tells you it is time to check up.

Once you are through adjusting the valves, make a last minute check of the pushrods. If one of them has bent and twirls instead of spinning straight, back up and trace the problem—it's a last minute warning sign that saved your engine before start-up. The source is

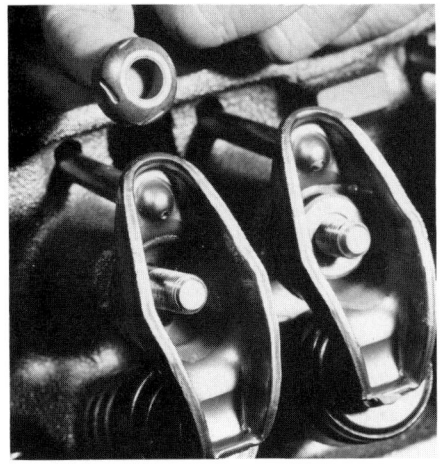

The ball socket and rocker need to have some moly rubbed into the surface for initial break-in. Locking type valve adjusters are preferred.

Each pair of Jesel rocker arms pivots on a common shaft with a support that bolts directly to the head. The rocker offset allows the pushrods to come closer together and makes room for much wider ports.

A button, piloted in the cam gear bore and held by the retainer, keeps the cam from floating out.

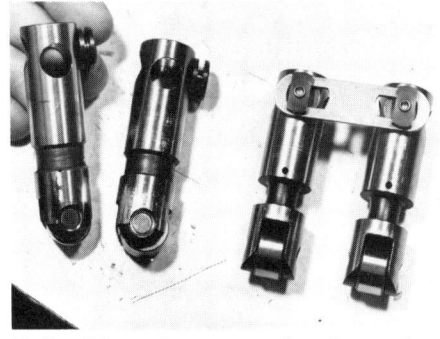

Roller lifters do wear at the pins and must be checked for looseness. You also need to check the guide bar and inspect the cam wear pattern.

most likely coil bind or a spring retainer hitting a valve guide. Fix it before putting it in the car. Final inspection: use a flashlight and a mirror to check that the rockers clear the valve spring retainers and that the pushrods do not bind up against the rockers. A hang-up between the pushrod and the head is a problem source and so is anything else that slows down the valve's return to a seat. These are all little things that have caused big failures for name racers at moneyed events.

PRE-OILING

Pre-oiling fills the oil galleries, pre-lubes the bearing, gets oil into the valve train, and most of all, it acts as a visual check that everything in the engine is OK for a startup. The oil filter is pre-filled and installed. You can use a cut down distributor and drive the shaft with a half inch drill or you can use the readily available pre-oiling shaft from

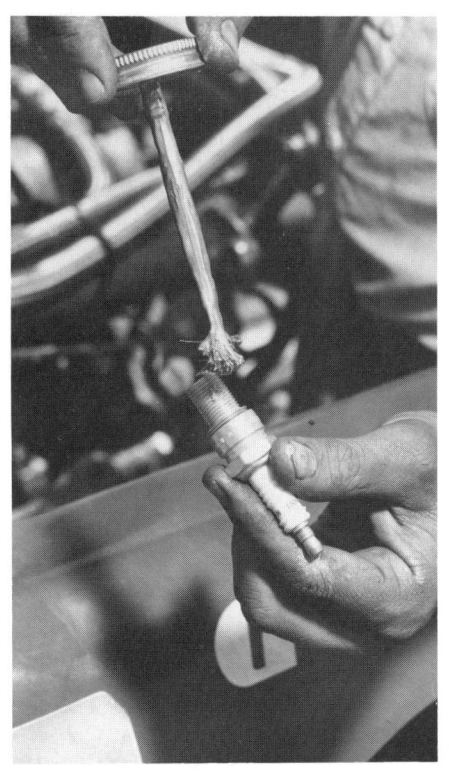

Use Anti-Seize on the spark plug threads, particularly with aluminum heads, to prevent metal pickup.

Mr. Gasket. Keep turning the engine as you pre-oil it and make sure that oil comes through with each pushrod. Always pre-fill and prime the oil pump as you install it. You can feed primer oil through the pressure pickup opening. A pressure pick up and a test light or gauge tells you when pressure starts coming up.

A roller lifter must ride square with the cam. Re-machining lifter bores on a mill improves the cam and roller lifter life and gains power by putting the lifter bores at the correct degrees. Fixtures are available from B-H-J to do this.

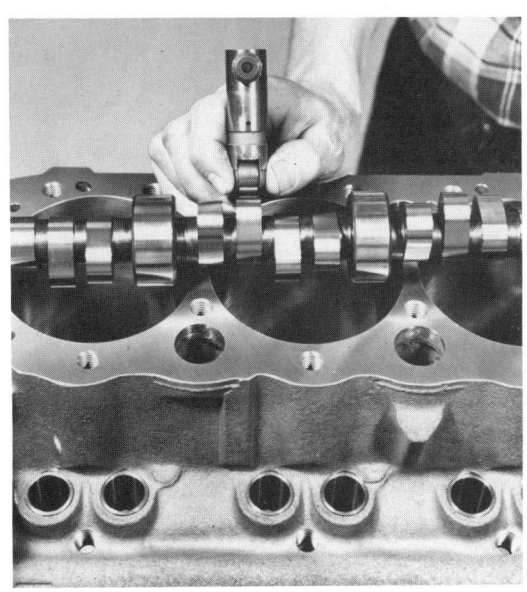

TOP END ASSEMBLY • 71

INTAKE TECH

To pick up torque you need extra runner length and plenum volume. Always test with different height spacers to see if it improves your particular manifold and carburetor application. Lab Machine's John Blackman is installing a 300-36 Holley on our dyno engine.

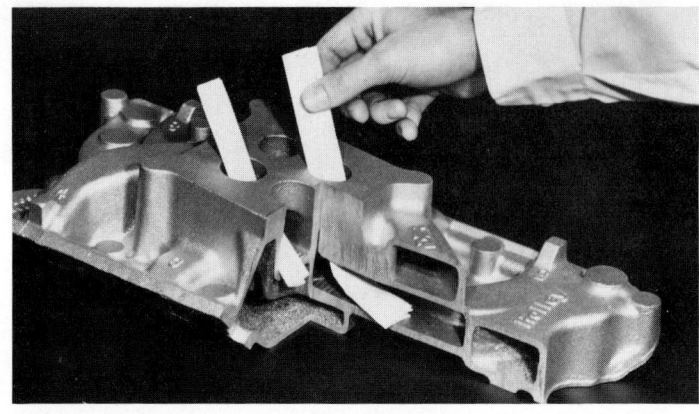

In a dual plane manifold, flow takes place along two levels, each one feeding two outer cylinders on one bank and two inner cylinders on the opposite bank.

We have seen the tall 300-36 pick up 50 lb.-ft. of torque over a single plane in the under 3500 rpm range. Above that, the single plane Street Dominator takes over.

If anything looks simple, beware. How can an intake manifold with no moving parts have so much effect on the engine, when it is just a way of connecting a carburetor to the ports in the heads? Possibly there is a lot more to it and it doesn't necessarily end with bolting on the manifold.

Involved are things such as the engine's rpm range, intake runner length, plenum volume, fuel distribution, and air distribution, plus the general manifold layout, be it single or dual plane. On top of that we start playing around with plenum volume and jetting as well as pump shot. All of the above, is involved in selecting the manifold and tuning it to a particular engine application.

Holley V.P. of engineering, Bob Miller who designed all of the Holley manifolds and Holley engineer, Jim Guibord, who has charge of intake manifold testing and development have both contributed extensively to this section and so have a number of engine builders.

People latch on to catch words like "reversions" and the thinking process stops. We're simply going to go back to basics without buzz words. We are on the exhaust stroke. The exhaust valve is open at TDC, the piston is at a near standstill and the intake has opened. It will take it a little time before the piston actively moves down. You would think that flow is at a standstill, but no.

If the exhaust is at a higher pressure than the intake, a certain amount of exhaust is going to make its way back into the cylinder. If you are running at very low throttle opening and the intake manifold vacuum is high, you can build all the dams in the world and the exhaust is going to go back into the cylinder, to occupy some volume. At full throttle with the right combination of cam, exhaust, intake and rpm, you can get exhaust scavenging and pull some of the intake mixture out of the intake and into the exhaust. Or you may have so much exhaust back pressure that it overwhelms the intake no matter how you tune it. Since most of us don't have a little man at each port, to check whether exhaust or intake has won the reversion battle, we go back to the old fashioned way of asking the engine which of many changes it liked best, or we turn to the reversion gurus and let them make a few more sales. The engineers at CPC, Chevrolet, and GM Research use dynamic pressure probes, look at pressure waves on scopes, sample exhaust mixtures, do computer analysis but when it is all done, go back to basics, like straightening out fuel distribution BEFORE you develop the rest of the engine.

If we tune the cam and intake manifold to resonate at 2500 or 3500 rpm, that is where it will help bump up the

72 • INTAKE TECH

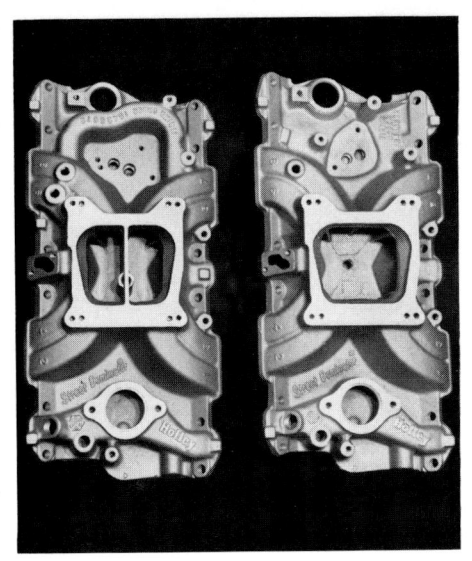

The 300-28Z has a rear runner joining the two halves of the manifold as a balance passage and the plenum has a center wall. It is extremely active and helps boost torque.

torque curve. You may tune your combination for 6000 or 8000 rpm and it will help the horsepower and torque in that range. However, what you do at one end can have the opposite effect at the other end.

PRESSURE PULSES

We sent our little scout into the engine, down the runner and into the cylinder head port. The report comes back on the squawk box that the intake valve just slammed on his nose and he can't get in—traffic is at a standstill. Next the valve just opens up and exhaust is climbing back up the intake port. Now the piston has finally started moving down and intake traffic is flowing into the cylinder.

Actually, the piston has started moving so fast that it pulls in more than the port and intake manifold can supply. Suction builds up in the cylinder, and air velocity in the runner increases. Our little scout is comfortably hidden in a pocket behind the valve guide and radios that the valve is slamming shut but air is still trying to get in.

The energy stored up in the column of air fuel mixture goes up as the square of the velocity. Meanwhile the piston reaches its maximum speed and starts slowing down because it must come to a stop at the bottom of the stroke. While the piston is not applying any extra suction, the column of air has enough energy to keep on ramming in. We take advantage of this by closing the intake valve well after bottom dead center (BDC). The upward piston motion is very small because crank and

The underside of a 300-36 manifold reveals the port layout with a long central runner feeding crossrunners. Manifold air flow is designed to keep the system active and to prevent fuel from dropping out. Note the exhaust heat riser.

connecting rod geometry makes it dwell longer at the bottom than at the top. There is a tradeoff between the piston moving up and trying to push air out of the cylinder, and the ram action at the intake manifold runner due to inertia, which is charging the cylinder.

A second action super-imposed on the first one, involves pressure pulses. The piston motion started a pressure pulse down the tuned length of the intake manifold. That pulse travels back and forth at the speed of sound.

Runner length will have a considerable effect on the pulse travel time. It will tune the arrival of a desirable pressure pulse as the valve is closing, but only for a limited rpm band. With a larger cross section area you may need a longer runner to stay tuned for a particular rpm. Each runner acts as a tuned organ pipe and sings at one or more specific frequencies. Ideally, the pressure pulse traveling down the runner is trapped inside the cylinder just as the intake valve is closing.

DUAL PLANE MANIFOLDS

With a V8, the intake manifold sees four intake pulses per crank turn. You can either keep them all together at a common central location or you can split them up. A dual plane Holley Contender manifold has two separate levels. Each of them services the two inside cylinders on one side and the two outside cylinders at the opposite bank. Together, the runners form the letter H with the cross bar of the H picking up from the primary and sec-

ondary on the one side of the carburetor. The upper and lower levels operate the same way and are just mirror images of each other.

Thanks to the dual plane Contender design you get strong pulses at each level, spaced 180 crank degrees apart. Having stout distinct pulses gains extra torque and better cylinder charging. It also makes the car more driveable.

There is a wide choice of Contenders, all tailored to a particular engine and power level. For instance, a 300-38 is great on a street Chevy with a stock 305 motor and gears. Holley dyno tests show that this manifold, bolted to a bone stock engine, adds 25 more horsepower at 4400 rpm. You definitely feel it in the car. It is designed to accommodate either a Square Bore or a Spread Bore Holley and Rochester throttle plate layout.

The next move up is a 300-36 which you can run with anything from a stock Q-jet to the newest Holley

The inlet and areas of the individual ports are carefully balanced out for even fuel distribution. The same manifold, plus an adapter will carry either a Square Bore or a Spread-Bore.

Holleys Contender at the left will accept a Spread-Bore or the original Q-Jet or the newest 4011 type Holley. Taller 300-36 at the right is set up for Square-Bore Holleys.

INTAKE TECH • 73

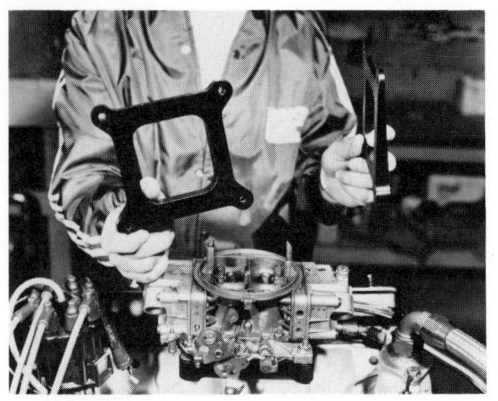

Spacers add plenum volume and change the pulsing action under the carburetor. Here, engine builder, Ray Baker is showing a wedge shaped spacer designed to level the carburetor on the engine for banked oval track use.

The engine on this Holley Carburetor dyno is ready for more pulls.

model 4010 or a Holley 3310 with vacuum operated secondaries. It is a tall manifold with ample port flow and we have had very good luck with it on the dyno and on the street. It is the manifold of choice if you have a strong running 350. The horsepower increase will depend on what you have in the engine. It can mean moving up from 160 to 200 horsepower but has also produced 360 horsepower at 5500 rpm with no strain. Perhaps not a record setter at national events but it is definitely a high class street torque builder.

SINGLE PLANE MANIFOLDS

With the single plane manifold all of the runners are on the same level and they all open up into a common plenum. The plenum volume is not just in the middle but also the volume of the other seven runners. In other words, each runner, in its operation, is looking at the other seven.

The single plane Holley 300-25 Strip Dominator or its big block equivalents, the 300-4 and 300-5 oval port and rectangular port manifold, are horsepower makers. Torque comes in at a higher rpm than a dual plane and they positively make more power at the top end while the 300-36 shines at the mid-range. There are also street versions such as the popular 300-1 designed for earlier 57 to 72 engines and a later 300-19 small block manifold and a low profile 300-3 big block.

Holley's latest single plane manifold works on Chevy small blocks with the Pontiac style high port heads. The part number is 300-41 and it bolts up to the tall Brodix made Pontiac style heads. It has both a tall floor and longer runners. The runners have been lengthened, which brings up the floor and the plenum.

MANIFOLD TUNING

The fact that you bolted on a new manifold and are testing it, is a step in the right direction but now you can post a few extra gains by tuning it. The manifold thrives on pulses which sometimes get to be a little too much for the carburetor. As a result, an extra spacer, anything ranging from 1/2 inch to 2 inches provides additional volume and height under the carburetor and smoothes things down a bit. For instance, an open spacer added to a 300-36 creates extra volume and cuts down on top end enrichment due to pulsations. You have to test the plenum on the car or the dyno to find out whether it helps or hinders.

As plenum size increases, former Holley engineer Mark Campbell from Automotive Performance in Monaca, Pennsylvania suggests adding or reducing pump shot—there is more air volume to accommodate with fuel and velocity can drop a little. The large plenum and the reduction in pulsations may call for larger jets.

The carburetor sees pulses going through it in both directions—down and also back up. With a lot of resonance through the carburetor you can see a fuel stand off—a fuel fog above the carburetor. This can create excess enrichment. Adding the spacer cuts down on pulsations, reduces the artificially high signal at the boosters and may call for going larger on the jets and the combination can well help economy.

Ex-Chevy Special Products engineer Bill Howell doesn't even begin to work on horsepower until fuel distribution is straightened out. There are so many variables in an engine that air and fuel don't necessarily arrive in predicted amounts at each cylinder. This shows

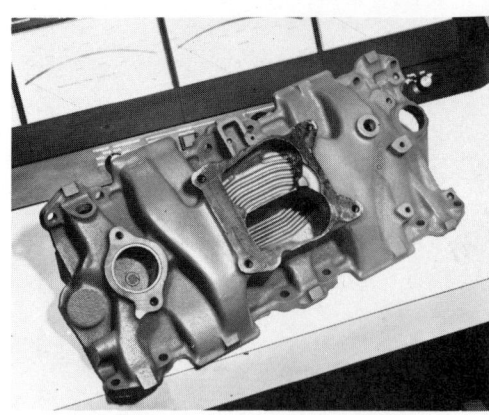

If class rules require it, a considerable amount of work can be done with the original cast intake manifold. Here is a Brzezinski reworked manifold with a substantially opened plenum.

up on the dyno as differences in the exhaust temperatures at each header pipe. The hot running cylinders are normally lean. If you redirect the fuel to the lean cylinder so that it can make full power and not run hot, you don't have to overenrich the other cylinders and the engine makes better power in addition to living longer. A little porting in one runner, a little work on the spacer, or a richer jet in one corner of the carburetor provide the correction. Simple things like tongue depressors epoxied in the manifold floor redirect wet flow and keep cylinders from going rich.

2 X 4 BARRELS

If you are running 2 X 4 barrels in a street application it pays to use a pair of Holley carbs with vacuum operated secondaries for better throttle response. To get the secondaries to open in unison, convert to P/N 20-73 secondary vacuum balance kit. It includes the Holley quick change black secondary covers with balance tubes. This guarantees equal vacuum and simultaneous secondary opening of both carburetors.

The throttle linkage is a little harder to set up; you can either use a common link at the sides or the better common shaft with link rods going to each carburetor. Keep the idle adjustments even.

The Holley 2X4 barrel Pro Dominator is a 300-23 for use with 4500 Dominator carburetors and 300-24 for 4150's (on S.B. Chevys), If you are running 4150's and want to switch to Dominators, just get the 300-201 top which is a bolt-on. The 4150 top carries Holley's 300-202 part number. Like we said, manifolds are great for bolt on torque and power—pick the one for your engine rpm range.

74 • INTAKE TECH

INTAKE BOLT ON

It's easy to install a manifold—nothing to it. Yet it is even easier to create a source of air or oil leaks which then haunt your effort to tune the engine. Here, we are going to seal up the leaks before they begin by going through a little check list—most of which you already know.

Check the intake bolt holes in the head and blow them out clean. If need be, use a bottoming tap—one which does not have a lead in tip or taper—and clean out the end threads. Now the bolts will not jam against the threads without tightening and give a false torque reading.

Always check intake bolt length—you don't want bolts long enough to touch pushrods. The bolt heads need installation clearance, which includes room for a socket. If need be, remove some material from the intake manifold with a die grinder. You may also want to use 12 point cap screws which accept a smaller socket or Allen bolts which use an internal hex.

CAUTION:

On a stock intake manifold, remove any cover used for heat shielding under the exhaust cross over—it will trap stuff a solvent tank will never clean off. Drill and tap at the drive rivet locations and reinstall the shielding. Use lock tabs or anaerobic sealer to keep the cap screws from coming out or reinstall the drive rivets.

To insure good intake manifold sealing, enough material is needed around each port. Use the intake manifold gasket as a checking template to see that sealing width is available at the heads and at the intake. If material is lacking for clamping the gasket, use a weld buildup on the intake manifold. Generally, leaks on the underside of the manifold oil up the plugs, cause excess oil consumption, and affect the idle. Leaks from on top of the manifold can sand the engine (suck dirt into it). The earmarks of an air leak are high idle speed—can't idle it down and erratic response to idle adjustment. When solvent is sprayed against a leak area, idle speed increases. Solvent can also cause a nice fire...

A manifold is a wedge which can ride up or down depending on the space between the heads. Milling the block or the heads changes the intake manifold spacing and the manifold is now higher in relation to the heads. To get it back down where it belongs, either mill the sides of the heads or the manifold flanges. If the manifold sits

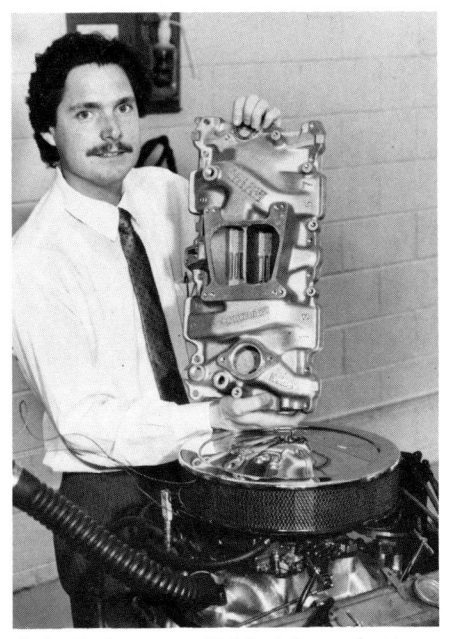

A dual plane manifold picks up bottom end and mid-range torque. Being able to run on the small primaries of a Spread-Bore adds fuel economy.

Reinstall the carburetor and the distributor. Reconnect the thermostat housing and all the vacuum lines.

During dyno test manifold changes, the ports and the tappet valley are protected against dirt. Scraping insures clean sealing surfaces.

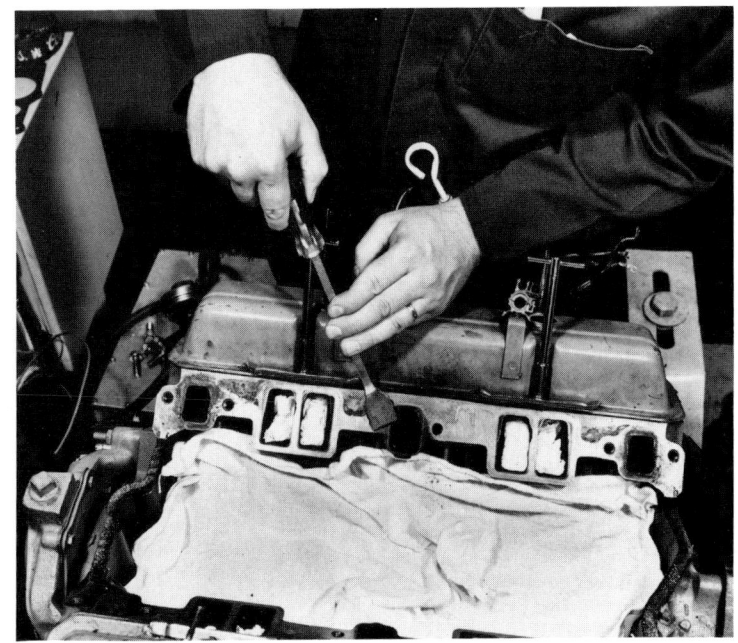

With the manifold seated on two flange gaskets, check the space between the block rail and the manifold. There should be enough clearance for a rubber cork gasket or a silicone bead.

too far down, you can sometimes save the day with double thick Fel-Pro intake gaskets or you can make up and install spacers. The angle to which heads and manifolds are cut may not match exactly. Do check the compressed thickness of the gasket along the top and bottom by miking it. If it is not the same, check the machining angles on the heads and the manifolds. A manifold can distort under tightening.

SEALING THE END RAILS

The intake manifold must seal against the front and rear block rails. When you shave the manifolds or the heads to match ports and lower the manifold, the front and rear block rails need checking and possibly cutting. Install the two flange gaskets on the heads, place the manifold on them, then use a feeler gauge stack to check the spacing at the ends. You need enough room for a gasket or for a bead of silicone. The bead of silicone should be applied in one continuous stretch from head to head. When you install the intake manifold, the silicone will squash a bit and overhang both sides of the rail, forming a saddle.

Rubber gaskets are not our favorite. Cork can pack down, while rubber cannot-you can squeeze it together but the total volume can't change. Being incompressible they tend to squirt out. Fel-Pro cork-rubber gaskets come through with a coat of adhesive, protected by a peel-off strip. Wiping the surfaces with paint thinner allows the adhesive to work. You can change the height of a cork-rubber end seal by 15—60 per cent; rubber end seals can only stand a 15—30 per cent height change. Less causes leaks and more can fracture the seal or cause it squirt out. As you can see the rubber is less tolerant of gap spacing. A third way is to simply lay a bead of silicone across the end rails. Make the bead just high enough so that when the manifold is tightened down, the bead forms a saddle over the end rails. Don't smear those beads or flatten them with your finger—make them even by controlling the pressure on the handle and the speed with which you pull the cartridge across the rail.

Cement the intake port gaskets to the heads so that the bolt holes are centered and lined up and the ports match. Put on the manifold without the end gaskets, let it sit for 15 to 20 minutes giving the cement a chance to set under pressure, then pull the manifold. This allows you to trim the gasket and clean up excess adhesive. Some builders use stake marks along the center divider walls to gain better side bite for the gasket in this thin unsupported area. Most engine builders also apply a light film of silicone around each port and around the coolant passage on the manifold side of the gasket for extra insurance. Now you can install the manifold and tighten up the bolts.

Bob Patzold of Prototype Engin-

The intake manifold gasket is cemented in place and centered at the bolt holes and ports.

76 • INTAKE BOLT ON

Apply a dab of silicone in the corners between the gaskets, to prevent oil leaks. The bead of silicone is applied in one shot with no interruptions. Note the PrintoSeal® beads at each port and coolant opening.

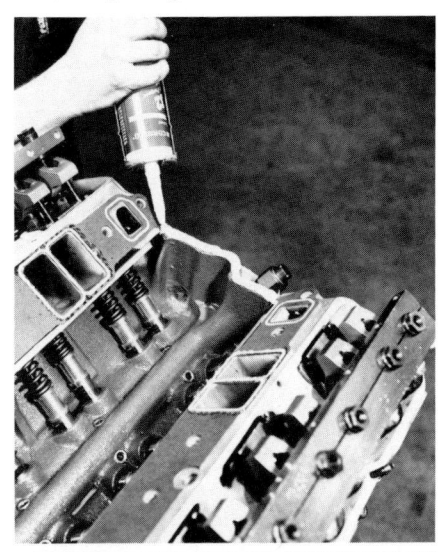

Lay in the intake manifold and center it. Without the tall Cork-Lam gaskets, the manifold will usually not clear the valve cover for installation.

Torque the intake manifold in a circular pattern from the center on out: preferably use a small universal to fit the center bolts.

eering explains that the intake manifold tends to slide around on a fresh silicone coat and so he goes through a few extra steps to guarantee alignment. He begins by leaving the intake manifold side of the gasket and the rails dry, lays on the manifold, and uses a flashlight at the ports and the bolt holes to insure alignment. The manifold is held in place and the gasket is marked with a ball-point pen following the outline of the casting. This gives you a quick line-up on the outside of the manifold as it is tightened.

Many engine builders, like Earl Gaerte, apply an extra coat of silicone along the top of the intake manifold gasket sealing area, on the theory that if there is a leak it will tend to draw in the silicone and self seal, instead of sanding the engine.

On a street engine always use a thermostat in the coolant outlet housing and make it a point to know its temperature rating. The thermostat acts both as a temperature control and a restrictor and you need both. On a race engine you can use a Moroso housing and restrictor or similar. Running the engine cooler helps spark knock.

If you are installing a stock manifold, read the warning sections in Engine Prep. Never beadblast an intake manifold, the beads have a way of finding their way into the engine no matter how carefully you clean the manifold. The exhaust crossover normally carbons up and seals itself off. Use a die grinder and wire brushes to restore flow.

INTAKE BOLT ON • 77

THE GREAT COVER-UP

The die cast cover in the center is more rigid. Note the knockout plugs and the bead to which you can heliarc. Rubber plugs with a lip can be inserted into the punched out hole. You have a choice of stamped steel, die cast, or stamped aluminum covers.

Tall acorn nuts make it easier to pop the cover. Cork laminated double gaskets are tall enough so that the intake manifold can be removed with the valve covers on.

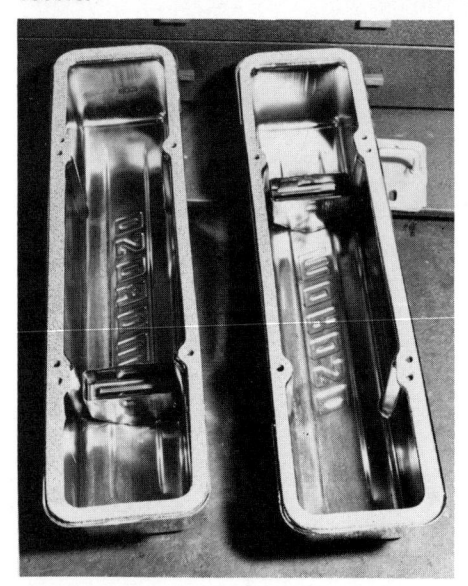

Cement the gasket to the cover, oil the underside, and you can reuse the same gasket many times.

That's what happens when stock rocker covers are installed backwards—they get dimpled, holes break out, and the engine gets oiled.

For circle track use, here is a die cast aluminum cover with a pair of TIG welded stacks and shielded breathers.

Oil impregnated breathers provide excellent dirt protection. Here is a series of different types of plug-ins ranging from a clamp on to fit a stack, to a rubber insert plug-in. Rubber plugs for PCV valves are also available and so are plug off caps.

You have completed your engine, bolted in all the major components, and are ready to install a pair of valve covers. A job so simple that absolutely anyone can do. If it is so easy, how come all the strange street engine leaks? Also, which one of the covers do you pick from the many catalog pages and how do you connect the breathers—50 other questions also come to mind. We took the easy way out on our last trip through Connecticut, spotted the Moroso building from I-95, and swung off to visit with Bob Rinaldi and Tom Abbott.

First job for a valve cover is to clear

78 • THE GREAT COVER UP

all the accumulated goodies. Where a street motor with bathtub rockers accepts low profile stamped steel covers, a racing cover is designed to clear roller tip rockers and screw in studs. Add a stud girdle and you really need the tall valve cover.

An engine is a thing of pride and beauty, so cosmetics have brought forth aluminum covers anodized in red, blue and black or stamped steel ones either chromed or ready to paint in the engine color scheme. Add to this die cast covers which lend themselves to ribs, radiused contours, and the more solid look.

SEALING UP

The one thing a rocker cover should do is prevent leaks, in spite of being flooded with oil above the gasket line. Technically, you want a rigid cover with an unbendable flange which will never leak. However, you also want it lighter than a brick and affordable, which calls for a thin steel or aluminum stamping. It is stiffened by an outside lip which also positions the gasket. The cap screws at the outer lip are distinctly insufficient because the gasket gives, the cover yields, and you get leaks. Now we add load spreaders, steel bars with a rounded pad for the cap screw and V-shaped outboard pressure points. Works great, just like using twice as many bolts. As Murphy's law would have it, the small block takes the wide spreaders and the big block uses the small ones. Load spreaders are built with just the right amount of flex, should not be overtightened, and are very forgiving.

Die cast covers have a more rigid lip and gasket support and do not need load spreaders. However, if you insist, do a little grinding and fitting to make the load spreader fit, and together they make a very good combination. If your assistant in the pits is a big dude who does not know his own strength, load spreaders are great insurance against leaks. Just explain that load spreader should not crush down in the center—that defeats the purpose.

STUDS, T-HANDLES, AND ACORNS

The best way to handle problems with cap crews, or stripped threads in the castings, or out of position gaskets, is to throw out the cap screws. Install a set of valve cover studs—Moroso, of course—use red Prolock II anaerobic sealer and they stay in place, and speed up assembly at the track.

On a drag race motor, T-handled wing nuts are great and save time. For a distance engine, oval track or street, where time and vibrations loosen anything, you want Nylock nuts and should replace them often. A show engine looks best with T-wing nuts.

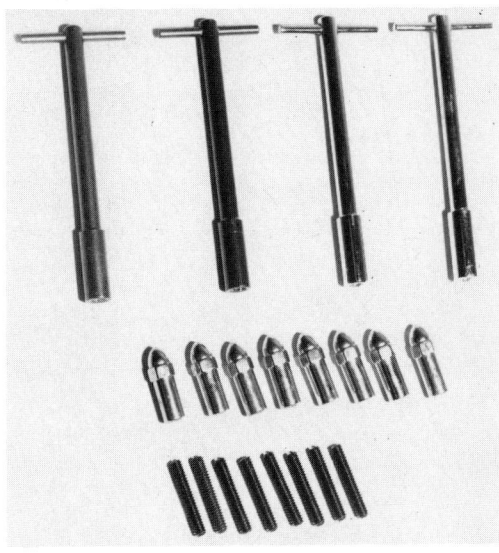

Baffle allows the oil to drain back without being thrown at the breather. Always check valve cover clearance without the gasket, then install the gasket.

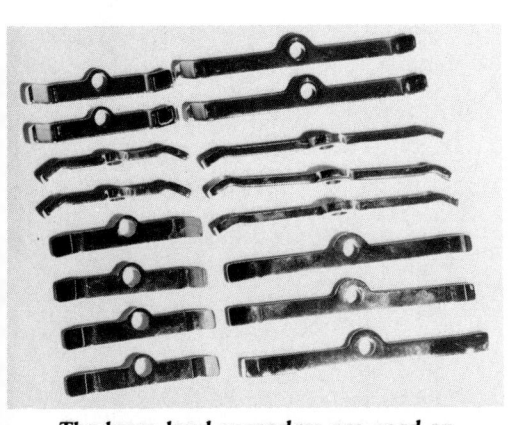

The large load spreaders are used on small blocks and vice versa. V-shaped part of the load spreader goes down and the design makes one bolt do the work of two. The center is NOT tightened against the cover.

These brackets, heliarced to the cover, allow a heat shield to be retained with Dzus quick release fasteners.

Acorn nuts at the right are a good decorative item which works on any street or drag engine. For oval track, use an elastic stop nut. The T-handle nuts are great for street or drag use, but they will not stay on as tight as acorn nuts.

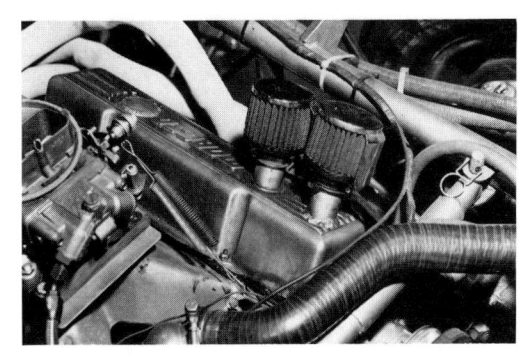

Here is a typical installation on the Bob Jennings' dyno car—two breathers clamped on welded stacks and a welded filler cover. Note the heat shield and heat spacer at the 750 Holley.

You have a wide choice of anodized covers, as well as chrome to make your rocker cover look its best. Ribs in the side of the cover provide wrench clearance and extra stiffness.

A valve cover needs a straight surface and on stock ones this means straightening it at the bolt holes with a drift and a hammer before installation. The gasket can be held in place with

THE GREAT COVER UP • 79

You will achieve the most leak-free valve cover installation by using studs. Also prevents stripped threads from frequent removals.

Fel-Cobond, 3M cement or Hi-Tack on the cover. The other side of the gasket is oiled so it will not stick to the head. Now you can pull the rocker cover repeatedly without damaging the gasket.

Ultra handy are double thickness rocker cover gaskets such as the Cork-Lam from Fel-Pro which come with a steel lamination in the middle and allow you to raise the cover and prevent a clearance problem.

VENTING AND BREATHING

Let's begin with a simple street cover which has no clearance problems with stock rockers. Here, there is ample room for a steel baffle that keeps oil from reaching the breather. A set of rubber grommets slip into the 1.220 Chevy sized holes stamped in the cover and now you can connect the breather and the PCV valve for the crankcase ventilation. Some of the tall valve covers are also available with a steel baffle, great when you don't have a stud girdle to contend with.

With a girdle in the way, you can use a rubber breather adapter which slips into the stamped steel or aluminum cover. It will accept a Moroso breather with a slip in extension or it can connect to a Vacupan system at the header collector.

The rubber adapters come in a variety of shapes which include one with an extension and a small slot—a simple, lightweight baffle that takes little room. On oil fills the rubber insert needs to be opened with a knife. There are also breather caps which push down into the rocker cover, same as on a stock Chevy, but with rubber that does not harden as quickly.

BREATHER STACKS

Oil slings out high up in the right-hand valve cover of an oval track machine and a stock breather location simply can't cope with the flood. The answer is to add tall breather stacks. Moroso makes a die cast cover with knock out plugs that can be punched out to accept breather stacks. Best way to insert them is to TIG weld the tubes in place. You can also use rubber grommets and slip in stacks.

HEAT BAFFLES

Headers or exhaust manifolds are a steady source of heat, which may be welcome on cold winter drive away, but do not help a bit on an Enduro car, or in hot weather racing. You can weld baffles directly to a rocker cover—extension wings to deflect the heat away from the carburetor. Moroso has the better idea, a pair of rocker cover brackets which can be welded to the sides. The baffles are retained by two Dzus fasteners apiece and the heat problem is solved. If you want to change plugs, pop the shields for full access.

One handy last suggestion—first check the cover clearance against the valve train without the gasket, then use the gasket and you know there is ample clearance.

As you can see, valve covers are easy, but there is a lot more to them than meets the eye.

The cover height depends on the valve train. The taller cover is a must for stud girdles.

80 • THE GREAT COVER UP

ENGINE HOW IT WORKS

1. Shucks, we just asked what's inside and the R.H.S. tool and die makers cut windows all over this small block with a Star Wars laser gun. There is no mystery inside the Chevy. It has been proven in 30 years of continuous racing development. Just in case your sweetheart asks you how it works, here are the pictures to show her the details. Surprisingly, the pros often make cutaways, watch the motions, and win races from having done this because it helps put things in perspective and develops new thinking.

2. The roller tipped rocker from Competition Cams is pushing down: now the valve and the spring retainer move down. You can see the cutaway guide and the open intake valve. The intake valve began opening just before the piston reached the top of its stroke; the piston is now on its way down for the intake stroke, with air and fuel coming into the cylinder. The exhaust valve is closed. You can see the piston pop-up, good for high compression, and the valve pockets which are a way to gain clearance between the piston and the valve. The cylinder is surrounded by a water jacket which cools it off.

3. The piston is moving down toward the bottom, and when it starts coming back up again, the intake valve will close, trapping the air fuel mixture in the cylinder. You'll find that the piston dwells extensively at TDC, speeds up, decelerates, and dwells longer at the bottom of the stroke (BDC) before speeding back up. The cam maker builds longer duration cams to help cram more air into the cylinder.

4. During compression both valves are closed. The piston is on the way up and the plug fires—well, you can't see the plug because it is in the piece that's cut away... The fire has to start early because it takes time to get it going— that's called spark advance. All of the burning must be completed as close to TDC as possible because that's when the volume above the piston is smallest.

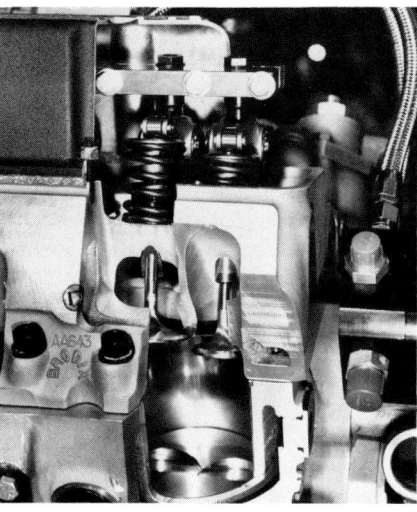

5. The piston travels down with both valves closed. Heat, generated by burning the air and fuel, expands the gases and produces pressure. You have to make heat to make power! Now the exhaust valve opens, venting the pressure in the cylinder. The piston will continue to move down, and then back up, chasing out the exhaust gases. The porting and cylinder head work done at R.H.S.—the folks who did all the milling on this cutaway engine—will contribute in a big way to air flow, cylinder filling, and the ability of the engine to burn more fuel and make more power.

6. The phase during which the intake and exhaust are both open is called overlap and contributes to improved scavenging—the original slug of high pressure exhaust may help pull out the last of the exhaust gases in the combustion chamber, while drawing in fresh intake mixture. Overlap can also have the exact opposite effect, leaving a bunch of exhaust gas trapped in the combustion chamber. That's where all the cam timing and manifold tuning science comes in, matching the cam to the engines rpm range.

THE GREAT COVER UP • 81

FINAL ASSEMBLY

When checking for a gear clearance, remove the solenoid housing and the spring. Now you can install the starter and move the drive against the flywheel teeth.

Your best bet is to use a high torque starter. It will last longer and will spin your engine faster. If a starter drags, the bearings are bad and it is time for replacement. When the amperage required by the starter—starter draw—is too low, you have a bad ground or bad connections at the battery. A Sears induction pickup ammeter is the best starter trouble shooting tool for the money.

Some people call it dressing the engine; we think of it as final assembly, all the things that need to be on the engine for it to run. You could hold off and assemble everything in the car but we are going to take advantage of being in a clean warm shop and check out most of the final assembly stuff right on the engine stand. Things like mounting the starter, running the spark plug wire harness or mounting exhaust manifolds are all best done ahead of time. Some items such as the distributor and carburetor should not be on the engine during installation in the car, but you can still do all your prefitting with them, things like checking that the plug wires will not interfere with the air cleaner. We'll start with the starter...

INSTALLING A STARTER

A starter is held in place by two special serrated bolts which pilot it. The bolts must be a snug fit to align the starter and its drive. Ordinary bolts will not work. There is a front starter bracket, extending between the starter and the block; leaving it out leads to starter failures. Starter shims can be used between the starter and the block to control clearance between the drive and the ring gear. If you like to replace starters frequently, don't check this clearance.

If the clearance is inadequate, the starter teeth bottom in the flywheel teeth, catch and either fail to engage or fail to release. When the starter gear clearance is too large, only the ends of the teeth are meshing, which applies a very high load to the teeth and causes breakage. If the clearance is excessive, the teeth instead of meshing properly hit each other and really break. Buy or make short starter shims for just the outside bolt position to tip the starter and reduce clearance.

Ron Charnes of ACC Rebuilders in Memphis, TN, gave us an effective way of checking starter clearance, by bringing the gears into mesh. Disconnect the starter solenoid housing, remove the spring, and the gear can now be freely moved. You need about a 0.020 clearance between the tip of the flywheel teeth and the bottom of the ring gear teeth. Install the starter, check that the

To accurately position a starter, you need a pair of serrated starter bolts. The shims are there to space the starter away from the gear—don't go by someone's advice, check the clearance on your assembly.

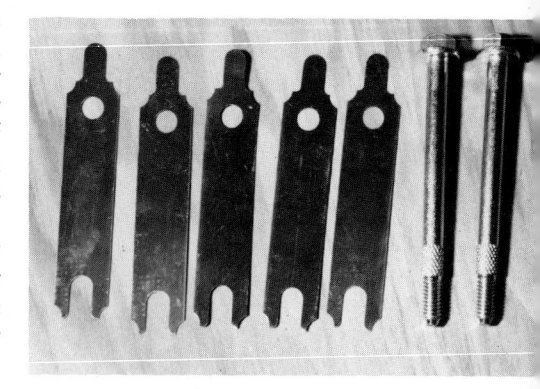

teeth move in smoothly, and set the clearance by adding or removing shims. On rare occasions, the block or starter housings are mismatched to the point that you need to mill the top of the starter housing to restore clearance. It is a lot easier to do all this with the engine on a stand in the shop before you install it in the car.

82 • FINAL ASSEMBLY

EXHAUST MANIFOLDS—HEADERS

An old fashioned cast iron exhaust manifold has many advantages—quiet and reasonably corrosion-proof. It also weighs too much and flows too little.

Each exhaust manifold has studs and nuts which retain the exhaust pipe flange. The studs are usually corroded and ready to break. Any muffler place can get them out by heating the manifold and turning out the studs, and so can you. If they break, don't write us—you have our sympathy. The only consolation is that they are easier to replace when the manifold is already off the engine. Studs should go in with anti-seize. Invest in some brass nuts to retain the manifold flange. If your parts man just ran out of them, order some because brass will not freeze on a steel stud. Brass or steel, you should double nut them. Do not leave out the small steel shields at the spark plugs—they help the plug wires and boots live longer. You can get fiberglass insulation

To check for even fuel distribution, the dyno heads are fitted with thermocouples: cylinder temperatures should be within 100–200 degrees of each other.

Here is a set of handy dyno headers at LAB Machine. For a valid test, check the headers and the exhaust system you will be running.

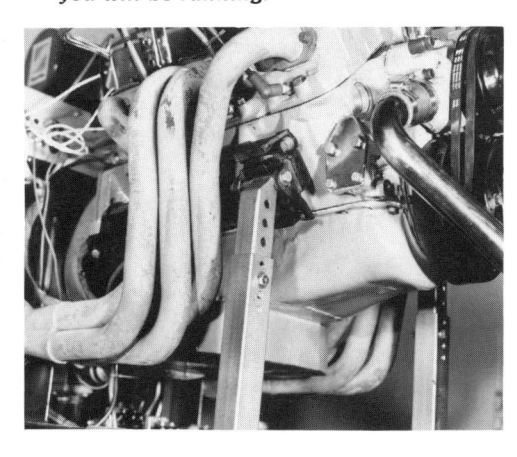

When you are fighting for torque, long header runners of reasonably equal length are essential. To kill excess torque on a slick track where you can't get bite, use "weed-burners."

in the form of sleeves and blankets to protect plug boots against heat. If you take time to coat the inside of the boot with MSD's heat resistant silicone they will be easier to pull off. Also, never remove those boots without a puller that snags them from underneath. It is an inexpensive plastic puller available at any speed shop or accessory store.

Exhaust manifolds collect a sizable amount of build-up which can be cut away with a die grinder. You can port match the exhaust manifold and the head to good advantage. Use new exhaust manifold bolts, washers, and anti-seize. The bolt holes in the head should be clean. Without anti-seize they tend to burn in place.

There are many levels of headers and price is a major consideration. A good header is aluminized, with enough material to protect it. A good header also has a thick flange that remains reasonably warp free. When you pick up a cheap header in the store, put a straight edge against it and find that there is nothing straight about it, don't expect it to seal. If you spend the money for a better quality header it will be more likely to live up to your expectations.

A good race header makes use of an adapter plate that bolts to the head with recessed cap screws (Moroso, Jere Stahl, Schoenfeld, Hooker etc.). This allows you to use a different bolt pattern at the headers with the bolts away from the tubes and eliminates struggling with the wrench next to a pipe. Do your header prefitting before putting the engine in the car. You will probably need to remove them again for the engine installation.

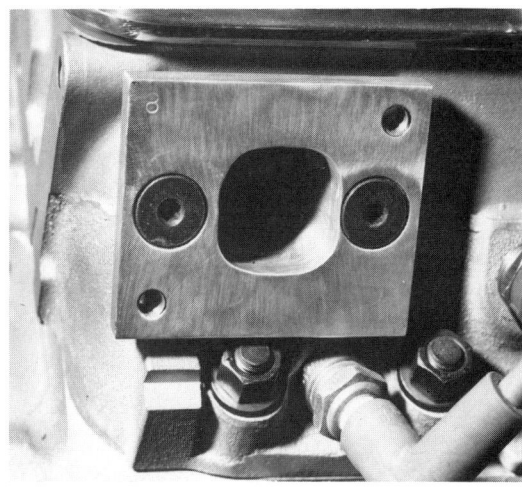

Moroso header adapter plate accepts recessed Allen screws and can be opened to the port. Now you have much more room for the header bolts, no interference at the pipes, and the engine can accept a better header.

Heat shields over the headers are popular on all of the Winston Cup cars.

The collector length and size has a considerable effect on torque and can be helpful in working out of a hole in the torque curve.

FINAL ASSEMBLY • 83

The fuel pump is activated by a push rod. For race use, Ray Baker makes up a roller tip push rod with a guide pin that keeps the roller square. Bill Mitchell sells a bronzed tipped push rod for use with billet cams while Holley offers a cover plate and an electric pump...

FUEL PUMP

If you plan to use an electric pump, simply install a cover to replace the stock mechanical pump. If you retain the mechanical one, install new gaskets inside and outside of the adapter plate.

A pushrod type follower rides against the cam eccentric and operates the pump. For billet roller cams, you can get bronze tipped follower rods or rods with a roller; the steel billet cams are not compatible with a hardened steel tip. A standard push rod works fine with cast cams on millions of engines. Make sure that none of the bolts at the front of the engine are long enough to contact the pump rod. A little grease holds the pump rod out of the way while you install the mechanical fuel pump.

Do make sure that you have at least a 3/8 fuel line, extending from the inside of the fuel tank to the pump. The tank portion is very important: if need be, remove the access plate of the fuel tank to run the proper pickup. As a rule of thumb, an engine consumes 0.5 pound of fuel per hour for every horsepower made. With a halfway decent 350 you need at least 150 pounds per hour fuel delivery. This must be delivered at 5-7 psi running pressure. To monitor this, see if the fuel pressure drops under full power as the car is going through the traps. Avoid right angle fittings because they cause a significant pressure drop. In running any fuel line play it safe—fuel causes large fires and you are in the car. Support all fuel lines with Adel clamps, keep them away from heat and from the drive shaft area. If need be, shield them in steel or aluminum tubing.

This bolt should not be long enough to touch or damage the pump rod.

SEAL UP THE ENGINE

You will normally not install the distributor and carburetor permanently before dropping the engine in the car, to keep them from getting banged up. Take time out to tape up all of the connections to the engine including the carburetor base, distributor mounting, coolant and oil connections, to prevent dirt entry. Modestly priced engine mounting stands are readily available for storage and transport.

INSTALLING A DISTRIBUTOR

The distributor is normally left out until the engine is installed in the car. It is a good way to keep it from being banged up against the firewall. If you will reuse the existing distributor cap and wiring harness, retain the No. 1 plug wire location and mark the rotor position at which it faces the No. 1 on the housing. The ignition system does not care where you locate the No. 1 wire on the cap as long as the distributor is timed to fire No. 1 at the right time. Do position the distributor so that there is room to swing it to set timing.

The damper and timing marks line up once every turn on compression and again on exhaust. This gives you a 50/50 chance of installing the distributor backwards. To identify that the cylinder is ready to fire, remove the spark plug, put your finger against the plug hole and you will feel a puff of air on compression. Another way is with the valve cover off—the exhaust will open and close, followed by the intake opening and closing and then the next firing point shows up before TDC, at the the timing mark.

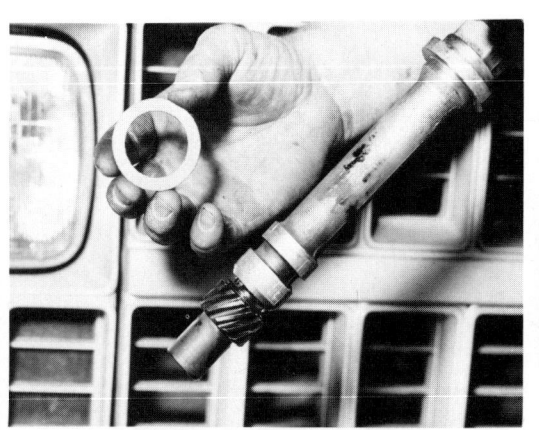

Oil pressure is routed around the distributor housing and this gasket is essential to prevent oil leaks. Check that the distributor is free to seat on the manifold without the gasket, which ensures adequate oil pump clearance when the gasket is installed. If the pump is bound up by the distributor pressing on it, you wipe out the pump and the cam gear.

Use a long screwdriver to align the oil pump drive with the distributor.

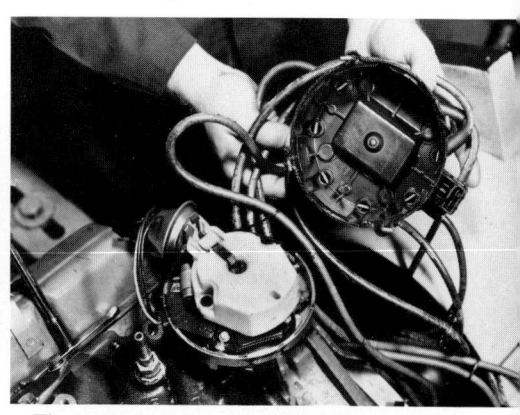

The rotor must point to the No. 1 cylinder post in the cap. As a quick check, turn the rotor against the spark advance return springs to see that it is free to move.

84 • FINAL ASSEMBLY

When you install a distributor, first check that the housing fits and can bottom against the intake manifold without a gasket. This tells you there is enough clearance between the distributor and the pump drive. Now, adding a gasket guarantees clearance. Without the clearance the distributor shaft bottoms against the oil pump and destroys it. A new gasket is essential because oil is routed around the distributor and the gasket is the external seal. When installed, the shoulder of the distributor must bottom so the clamp will lock up the timing. While setting timing, leave some slight drag at the clamp, so the distributor can be moved, but retains its position.

When you install the distributor, you'll find that the helix of the gear turns the rotor as you push down the distributor. If the rotor goes beyond its intended location, just pull out the distributor and move it by one tooth in the direction you need. Since the oil pump drive must line up with the distributor, reach in with a long screwdriver and line up the pump drive. Again, very important: when installed, the distributor housing shoulder has to bottom.

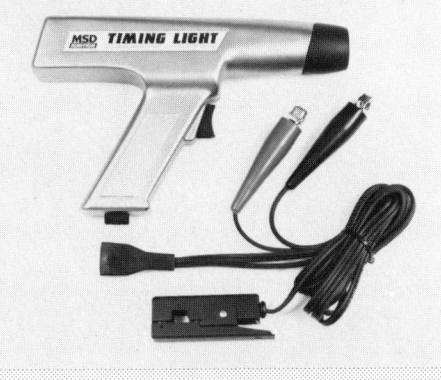

Conventional timing lights can induce excessive actual settings and fry your engine, because they have their own built in retard. Circuitry in MSD's racing timing light gets the timing right on and can stand the high powered output of the MSD box. Clamp on induction pick up has a steel housing instead of melt-down-plastic.

TIMING LIGHTS

Timing synchronizes the spark with the crankshaft and the piston position. Any time the distributor has been pulled and reinstalled it must be retimed. It is impossible to make an intelligent decision about engine tuning, if you don't get the basics right with new plugs and correct timing.

MSD's Jeff Gaul points out that not all timing lights are created equal. El cheapo timing lights that derive all of their power from the plug wires don't put out enough light to see by. You need a timing light that relies on the car battery for power. It should also have an induction pickup that clamps around the sparkplug wire. Clamp it to a straight wire section, away from other wires. Lacking an induction pickup you can use an adapter wire with a Tee that connects to the timing light.

OIL GALLEY

Oil Bleed Hole For Distributor Gear

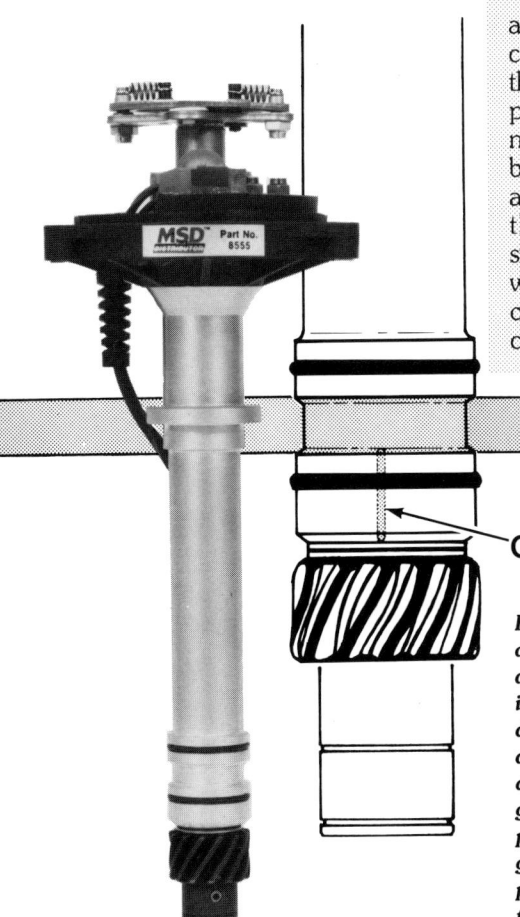

Here is a new approach to distributor and cam gear wear—just spray some oil and the problem goes away. An 0.030 inch oil drilling through lower part of the distributor boss sprays oil on the gear and lubes it. The new MSD billet aluminum housing also carries two grooves with special O-rings to seal oil pressure above and below the oil galley—it makes a noticeable oil pressure improvement. Oil pressure to the galley and groove can't escape at the distributor housing to block clearance.

Always recheck the ignition timing with a timing light after reinstalling a distributor. Hand-feel tells you when the cap has engaged the distributor housing.

Cheaper timing lights will disintegrate in front of your eyes when use with a high powered ignition—we have several like that in our collection. Some of the hi-tech ones with the knob for total timing read-out, will "drift" shamelessly, giving you ever changing readings as they warm up. MSD's timing light and the chrome Sears timing light, both with an induction pickup, give accurate results.

To get the most use out of a timing light, you need to be able to read total timing at different rpm and at different vacuum readings. To do so, degree the timing damper. You can clean the damper and cement a timing tape to it or mark the damper, to read to 40 degrees. Another way is to chisel in marks through the tape, which makes them more permanent. The mark you want to use should be painted white or yellow, to make it stand out. Always try to reach best torque with the least timing. More is not better. You need less timing with an open chamber and a flat top piston than with one filled with a pop-up.

Connect a tach to the engine, disconnect and plug the vacuum line to the distributor and see how much advance you have at idle. Run up the engine rpm to 1000, 1500 etc. till the spark stops advancing, usually at 3000 or below. Put the numbers into a little chart and you will know where you are at. The chart tells you the initial advance—how much advance there is at idle. It also tells you total advance. Now reconnect vacuum advance and see how much it adds. You can use a Mity Vac pump connected to the distributor vacuum to come up with more information.

FINAL ASSEMBLY • 85

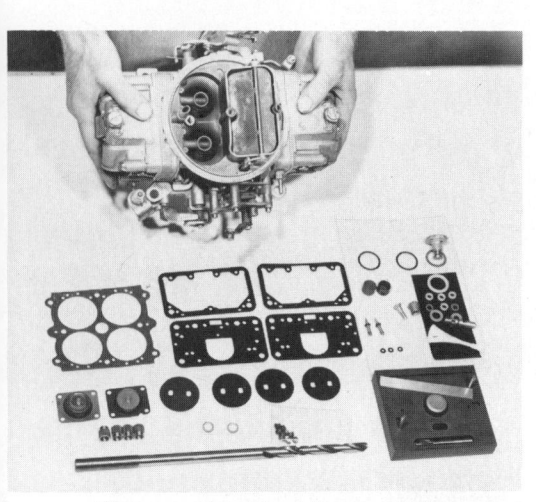

This complete Brad Urban kit from the Carburetor Shop includes a step drill and bushings to prevent throttle shaft looseness. The block in the right front helps straighten metering bodies and there is also a collection of useful nylon washers.

INSTALLING THE CARBURETOR

Your job will be easier if you tagged all the vacuum lines at disassembly. Also, buy enough hose of the right sizes to replace tired hoses. Before you install a new carburetor, check that the carburetor gasket fits the manifold and fits the base of the carburetor. It avoids spending a week fighting mystery vacuum leaks just to find an incorrect gasket.

With a performance engine spend the time to test different carburetor spacers and do some tuning. You should have a heat barrier between the carburetor and the manifold, but avoid the stack of soft gaskets—they are the direct cause of bent carburetor flanges. Tighten the carburetor nuts evenly and in small increments. You should feel a sharp increase in torque as the nut tightens, not a slow mushy torque rise as flanges bend.

If you plan to do frequent jetting work, spray the gaskets with lubricant to keep them from sticking. For good sealing, install the gasket as is.

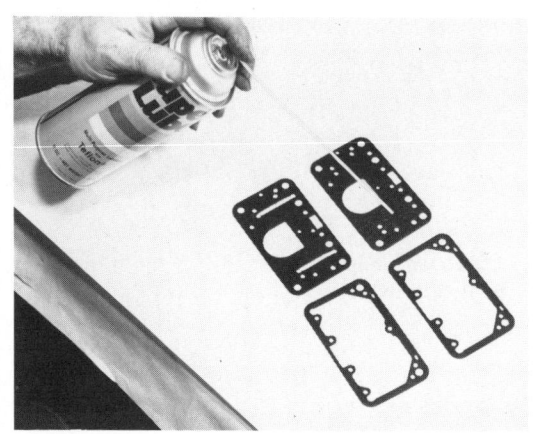

Make sure there is a large in line fuel filter to protect the carburetor and that the lines have been routed away from heat. Vacuum hoses need to press on snugly.

Connect the throttle and the transmission kick down. The throttle should be able to return to idle with the push of a finger with no help from the return spring. If you need a stiff return spring to get idle, the cable or the linkage is wrong and must be fixed and readjusted. You also need to reach wide open throttle when the pedal is on the floor. Things as dumb as too much matting under the accelerator pedal or incorrect throttle cable adjustment keep the engine from reaching full throttle and max power.

By the same token, do install a throttle stop under the pedal, so that you can't apply full leg force against the throttle linkage. Spend a little time on carburetor adjustment—it will net a noticeable improvement. Read *Supertuning Holley Carburetors*, it will save ET's or *cut* lap times by helping you smooth out the idle to main jet transition. *Eliminate* sag and bogs, by working out the best power valve and accelerator pump combinations.

There is no limit to the amount of work that you can put into a carburetor to make the car work better... or worse. If you don't make some mistakes, you also won't learn anything.

The throttle and kickdown cables should be adjusted. The cable needs a little slack at idle but must be able to open the throttle all the way. The engine will race if the throttle cable doesn't release all the way.

Tighten the four carburetor base nuts evenly to keep from bending the throttle assembly.

VALVE COVERS

Read the Valve Cover section. Just in case you are short of time, here are a few tips:

Have the load spreaders on hand. Use valve cover studs rather than screws, and retain with Pro-Lock or Loctite anaerobic sealer. The valve cover should fit so it does not touch the rocker arms or the stud girdle. Also check that it clears the intake manifold. In many instances, a Cork-Lam sandwich type gasket with a steel shim between two cork-rubber layers will raise the cover sufficiently to help intake clearance.

Use a heat insulator and gaskets between the manifold and your Holley. The negative side of the electric choke must be grounded and the positive side needs a supply from the ignition switch; a direct supply kills the battery.

86 • **FINAL ASSEMBLY**

START-UP

Pre-oiling is essential to your new engines well being. It allows the bearings, lifters, and valve train parts to be prelubed and it serves as a final warning if there is no oil pressure.

After a break-in, the valve train is pulled, the heads are retorqued, all bolts are retightened, and now you are ready for some serious tuning. Here, Ross Lombardi is putting the finishing touches on an alcohol injected motor.

Ideal place for a break-in is on the dyno where you can spot leaks, check pumps, retorque and test. Or, just take it out on the track and romp on it...

To win, get a good ignition system. Here, an MSD 6-T is combined with a Blaster-2 coil. Pigtail on the MSD 6-T allows the use of a Soft-Touch Rev Limiter.

Here is a guaranteed, quick, cool-down system that can be run from an auxiliary power generator. The two Grainger fans are mounted on a support plate that fits in front of the radiator. It is standard Winston Cup practice.

COOL IT

The stock radiator is grossly inadequate. A single row core just can't handle a muscle engines heat rejection rates.

Enduro, Bomber, Late Model, Hobby Stock, or Winston Cup—you are carrying fuel and need some safety insurance. Impeccable Winston Cup installation with fuel cell, check valves to hold the fuel in case of a flip, rollbar protection, and a carrier for the fuel cell—this is not the place to shortcut when you build your next car.

When you order a custom radiator, get a core with enough tubes; outlets can be positioned where you want them.

Finishing touches are being put on the core at Griffin Radiator in Townville, SC. You can order them in brass or aluminum.

FINAL ASSEMBLY • 87

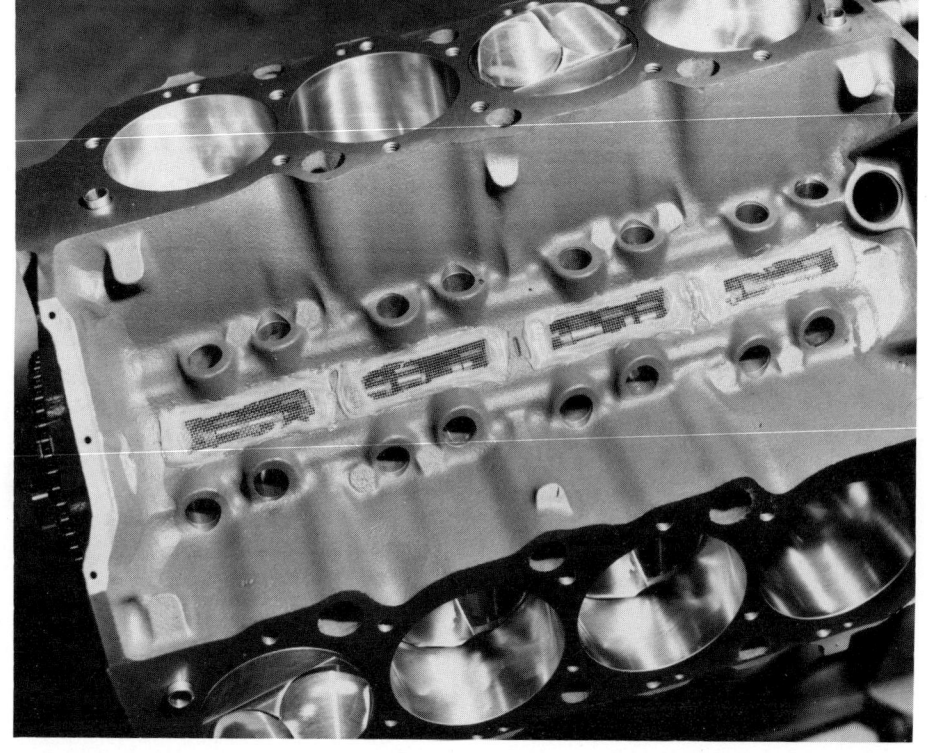

You are entitled to unlimited 638 cubic inch power from a Bill Mitchell motor. But 570 cubes is a good practical size..

MONSTER MOTOR

To go to the moon rapidly, contact the nearest NASA space center. If, on the other hand, you are seeking some serious ground speed achieved on less than the Federal budget, you will have more success by contacting Bill Mitchell on the subject of practical big inches. Big used to mean putting down the coin of the realm for a 454. By today's standards that is modest fare indeed, sort of an overgrown small block.

The range we're going to cover begins at 454, soon moves up to 490 and 511, then makes a short stop at the 540 cubic inch level. Having just escaped earth's gravitational pull, you can cruise onto 570 cubic inches which is on the high end of affordable practical cubes. In fact this fits both the street and the drag image. From this point on you can get a little more exotic with a 604. Then too, you can set the boring bar a tad larger to come up with a 638. That plus the NOX button is guaranteed to create a moving machine.

Under normal circumstances it is not a great trick to create a machine which will push out 1.5 horsepower per cube so you will be right up there nudging the 900–1000 mark on gasoline, plus whatever you plan to add in laughing gas. We're going to start this project by taking a look at the options from a stock block to a Bowtie and in a wide range of bores and cranks. The strange part of it all is that sticker shock will not set in until you start pushing for the stratospheric limits, beyond 570. Instead, you are dealing with less much money than is called for in a Pro/Stock engine and stand a chance of going faster than anything envisioned in class rules.

We are going to start at the Motel 6 level with a stock 454 big block and work up. Here, the limit is 509 cubic inches. The block better check out well before investing any money in it. Look for core shift by first seeing if the freeze out plugs are centered in their bosses. Also, check the cam bearing bores. If those are acceptable, the cylinder walls should be sonic tested. The largest stock block combination will increase the stroke from 4.00 to 4.25 with

Screens protect the bottom end from valve train pieces.

88 • BIG INCHES

BIG INCHES

crank and block clearancing, and will add 0.125 to the bore size, going from 4.25 to 4.375.

Still on an economy big inch kick you can take a logical step by investing in a tall deck truck or marine block. Here the block deck height increases from 9.80 to 10.2 inches. This is the distance measured from the center line of the crank to the top of the block. The extra 0.400 inch will give you added connecting rod room. Where the stock block accepts a 0.250 long rod, the tall deck block can use a 0.400 long rod. The longer rod allows more piston dwell at TDC, gains top end horsepower, and reduces the side loads against the cylinder wall.

BOWTIE BLOCK

If you plan to make full use of the mega torque derived from big inches, start on solid footing with a Bowtie block. It has more metal at the main bearing webs and thicker decks. Where stock bolt holes go right through, these are closed off at the bottom and have a better tie-in. The cylinder barrels are siamesed and you can bore to your heart's content. Bowties are available in both short and tall blocks and now that the tall deck Dart intake manifold is readily available, it has become the more practical of the two.

Strong as it may be, the Bowtie can stand a little reinforcement. Mitchell feels that Chevy should have filled the bottom of the cooling area with more iron to prevent water leakage and provide increased block stability. Lacking this, you can fill it with machine grout which takes a couple of days to set but is worth the wait; the other alternative is to fill it with two pounds of Devcon to a side. Drag motors are filled to within two inches of the top while a street motor needs filling to the bottom end of the freeze out plugs. This leaves more room for coolant while still strengthening the block.

BORE AND STROKE

Before diving into the details of each combination Frank Cerasuola, Bill Mitchell's head machinist, takes us through some of the mechanical bore and stroke limits. You can easily bore to

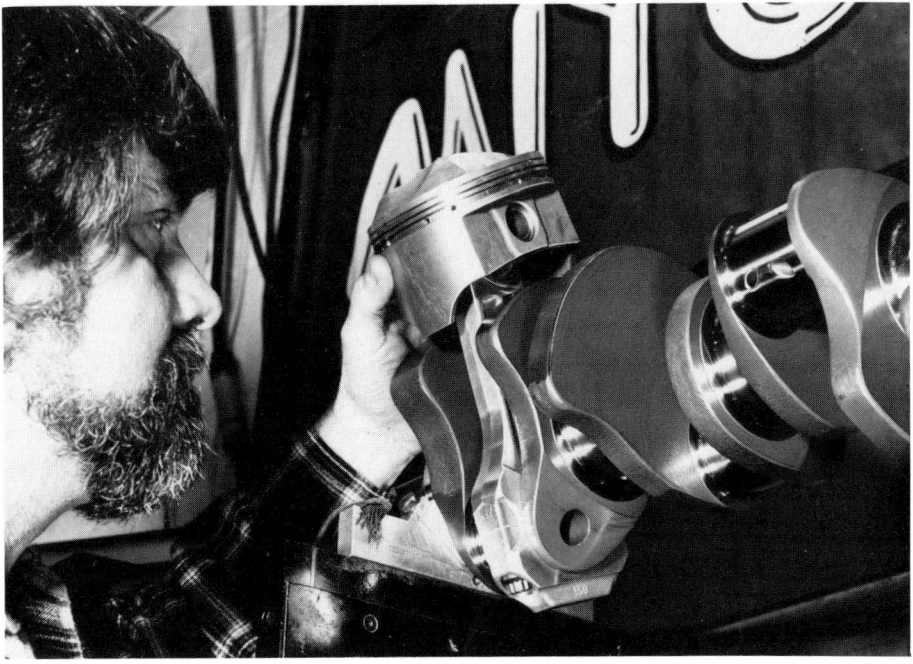

Big inches call for moving out from the stock 4.00 stroke to 4.25 and 4.50. One look at how far the rod journals are from the mains tells you that this is a serious BRC crank.

With the stroker you need a long rod to reduce rod angularity and to reach a higher rpm band. Now the wrist pin is higher, and closer to the cylinder head deck, so there is less room for rings. The long rod helps keep the bottom of the piston away from the crank.

The long stroke brings the piston further down into the bore and closer to the counterweights. BRC cuts their counterweights to an elliptical shape, using a smaller radius at BDC (bottom) for clearance and a larger radius towards the ends for extra metal.

BIG INCHES • 89

Use Devcon to fill the block to the low end of the freeze out plugs for a street engine, or to within two inches from the top for a drag machine. →

4.375 which is a standard 4.250 Chevrolet bore size plus 0.125. This takes the minimum penalty in bore size, retains very stout cylinder walls, and gains a fully rebuildable engine in case of a mishap. A 4.5 bore gains more displacement, leaves extra room around the chamber wall for flow, and has oversizes for a rebuild. All of the common 4.520 gaskets such as Fel-Pro fit and will not crush into the cylinder. The 4.5 bore is the best, and most practical. There are rebores for it from 4.530 all the way up to 4.625, some more readily available than others.

The ultimate bore is 4.600 but it comes with penalties. There is no margin in repairing cylinder walls and you are into copper head gaskets and O-rings. The other problem is getting rings. They are plentiful for a 4.5 bore but at 4.6, better check availability. Also,

A reamer machines the front and second counterweights from each end, making room for extra slugs of Mallory metal. ↓

There is the finished crank ready for balancing. Bob-weight depends on whether you use a steel or an aluminum rod. ↓

90 • BIG INCHES

Computer controlled balancer checks for the amount of unbalance independently at each end of the crank.

Since Mallory metal is heavier than steel, it increases the effectiveness of the counterweights. The slugs of Mallory metal are pressed in and then welded. It is important to put them in from the side so centrifugal force cannot throw them outward.

These 0.250 inch long Carrillo rods keep the piston skirt from colliding with the cam ground BRC counterweights.

a 4.625 bore tends to strike oil at the galleries leading to the cam bearings. Now the galleries need sleeves. It is good insurance with 4.5 inch bores and is definitely required with 4.625 bores.

You can easily take the first step" requires a bit more work. At 4.5 inches machining is required at the cylinder barrels and at the crank case rails, but nothing outrageous. Here aluminum rods are more of a problem than steel ones.

It does not pay to hand fit the block with a die grinder. Bill Mitchell simply puts it in a big mill and cuts away to a preset pattern for clearing the big Carrillo, BRC, or Childs & Albert rods. By the time you reach a 4.75 stroke there is no way to fit aluminum rods, and even with steel rods and a smaller base circle the cam becomes an obstacle. Now that we are done with the basics, let's take a look at large Bowtie block combinations that will wake up your machine.

540 CUBES

The 540 is an easy engine to build—you can achieve it with either a short or a tall deck block. Rod length is limited to plus 0.250 with a short deck and plus 0.400 with a tall block. You will need a piston with a 1.270 compression height from the center of the pin to the top. It has room for three rings and at 620 grams it is quite light. The crank counterweights are cam shaped rather than round, which adds clearance for the piston at BDC but still retains enough weight. You should know the rod length ahead of time so that the BRC crank is contoured accordingly.

572 AND 618 CUBIC INCHES

If you wish to break the estimated travel time record from earth to Space Station-4, a 572 square combination

BIG INCHES • 91

The Dart carries ample extra metal and lends itself to good port work. It begins with 360 cc ports and let's you take it from there.

Jesel belt drive allows the cam to be degreed or moved without pulling the damper.

with a 4.5 stroke and a 4.5 bore will do the job. It does require the tall block and the 0.400 rod. You now have more room at the top of the piston with a 1.395 compression height and the oil ring groove easily clears the pin area. The engines run best with Speed/Pro 1/16 moly–1/16 moly and -3/16 Speed/Pro SS50 oil rings. Bill suggests not using low tension oil rings on the big engine—"you'll never be short of torque and you preserve the clean air!..."

The no-lift portion of a cam is called the base circle—it is a convenient way of describing the physical size of the cam. If you installed a cam with a standard 1.100 base circle into a 572 big block the rods wouldn't clear. To gain the necessary rod clearance calls for a smaller 0.975 base circle cam. The rods themselves are not modified.

604 PLUS

A 2300 pound Top Sportsman door-slammer with a 604 makes its Pro/Stock counterpart look a bit limited. By the time you have made room within the crankcase for a 4.75 inch crank stroke with a 4340 chrome moly non-twisted forging you have definitely made your statement. The displacement is now 604 cubic inches, enough to make the Lenco more respectful. Should you wish to go totally wild, a 4.600 bore brings the displacement to 638.

As you go to 572 cubes and up, balancing the crank involves more than casual massaging. In addition to working out the crank design details to optimize the counterweights Bill Mitchell also proceeds to drill and ream through the two front counterweights as well as through the two rear ones, which doubles the rows of Mallory metal slugs. With the super heavy Mallory, the crank no longer requires external weights.

Your choice of steel versus aluminum rods is determined by many considerations. Obviously, the steel rod is more durable but it weighs more and costs twice as much as an aluminum rod. With the set of 0.400 long rods there is 150 grams per rod difference—that is nearly three pounds for the entire rotating assembly.

CAMS

The idea of the big motor is to have all sorts of torque, even if you go easy on the rpm. You will normally twist between 6500 and 7000 even though some go to 8000 rpm. The bigger the engine the more cam it can stand. With a boat you may be in a 265—272 degree range, while for drag racing you may need a 282—286 or a 286—294, all at a 0.050 lift. The valve lift will be in the 0.650 range. Billet roller cams are the only ones used. Naturally, you are into 165 pound seat pressures and 465 open pressures. You can use one in a boat or a street machine with a hydraulic cam. As rpm and the valve lifts increase, the spring pressures move up to 205 on the seat and 500 open.

Each car owner has his own preferences but the top end trend is toward the more rigid Jesel rocker shaft setup. It eliminates the stud girdle, cuts down on deflections, and stands still for 225 seat and 600 open Vascojet springs. Installation of the Jesel assembly involves machining the head and using a special fixture with a dummy rocker. Adjusting shims are used to get the rocker centered on the valve. Also popular is a Jesel timing drive with a split timing case cover which allows rapid cam center line changes.

Bill Mitchell supplies long pushrods in kit form. This enables you to fit any engine combination, cut the pushrod to size and press in the end.

HEADS UP

The Godzilla motor combination calls for heads to match. Fortunately both Brodix and Dart saw the large potential and both supply heads for the purpose. With the smaller 540 or with stock block 511's a 320 cc port is adequate. For 570 or 600 cubic inches you can upgrade to a Dart head with 360 cc ports. There is room within the

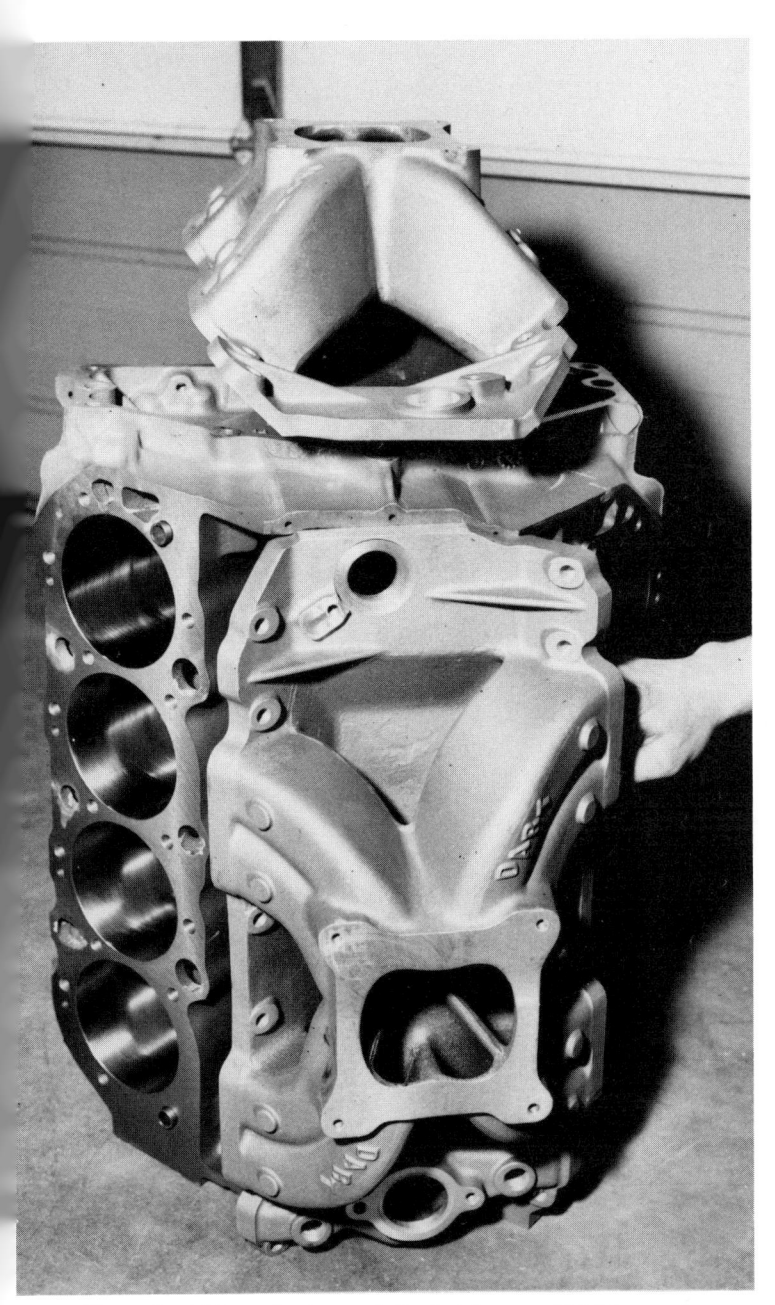

The rail width at the base of the intake manifold is greater on the tall block. Dart makes a manifold for both applications, with the same free flowing runners.

You can retain the stock distributor with either the stock or the truck block, because the corresponding Dart manifolds have a tall or a low mounting pad to accommodate it.

A stronger crank has a larger radius tying the journals to the cheeks. Here, the bearings are chucked in a holding collet and beveled to clear the massive crank radius.

Dart and Brodix heads to increase the cc as needed.

Dealing with Mitchell's Monster motors it is easy to find compression sneaking up on you. When you take that 454 out to 572 or 604 the difficulty is to find enough combustion chamber cc's. A stock open-chambered head measures 119 cc while the Brodix head comes in at 120 cc. Dart supplies a 138 cc chamber out of the box. If your engine combination calls for it, a 0.100 angle mill can whittle the combustion chambers down to 90 cc.

Most gas machines favor a 13 to 13.5 compression while some alcohol tractor pullers run at a 14:1 compression. A boat is usually content with a 9.5:1 with a flat top piston.

INTAKE MANIFOLDS

We saved the best for last. There used to be a time when builders of tall blocks struggled with spacer plates to accommodate standard width intake manifolds. Now, Dart makes available two manifolds one for the standard block and the other for the tall deck; they have different heights and widths and form a direct bolt on. The distributor pads are set at different heights so that you can run either of them with a stock distributor.

When you compare the drive feel of a small buzz bomb low cube engine to the total torque available from 540 or 604 cubic inches it feels as though an extra engine was just bolted in. Big cubic inches will give you quite a head start if you want to go fast.

The smaller 0.975 base circle cam allows the rods to clear with a 4.250 or 4.5 inch stroke.

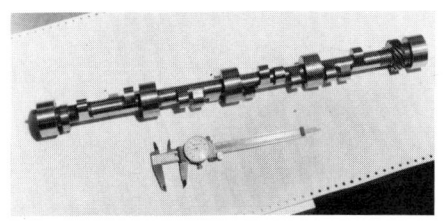

BIG INCHES • 93

457 SMALL BLOCK

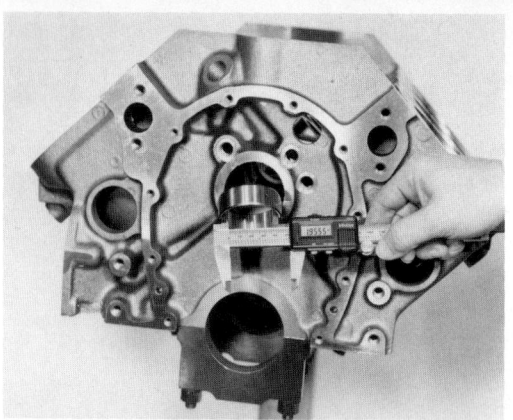

These cam bearings will not spin. Not only is the bearing larger but also the cam bearing oil grooves have been eliminated from the block. Here, the cam bearings get solid support.

The latest small block carries as an emblem a Rocket instead of a Bow Tie. It wears a GM badge and as such is available at all GM dealers - read Chevy. While sanctioning bodies will probably have a thousand rules on the subject, we expect it to become a popular Chevy part and a standard of the industry. When you want to build a big cube engine within a small block, the standard block basically runs out of metal. You can machine the block rails for rod clearance and you can also rework the sides of the rods but the net space limits are tight.

On the Rocket block you can easily bore to 4.200 leaving ample wall thickness. Oil pan rails have been spread 0.800 inch which leaves enough clearance to install a 4.125 crank with no problems. Also, the cam has been raised 0.400 inch so the rods clear. Building a 457 allows you to look a guy in the eye and say "I'm just runnin a lil ol'e small block..."

Now that we have your full attention, join us for a rapid trip to Lubbock Texas, the home of Scoggin and Dickey, the huge Chevy dealership with the hottest race and performance parts. Nick Fowler had the first of those Rocket blocks in stock and his "hobby shop" was already building one for oval track and another for what promises to be the hottest street machine, even by Texas standards. This, as you can see is a multipurpose piece with which you can achieve anything from a stout 350 to a conservative 400. With very little extra effort you get a 427 or a 457. Not bad for a small block with some extra metal in all the right places.

HARD NOSED

There are difference in cast iron and the one from which this block is made is harder stuff than either a street block or the Bow Tie.

The deck is massive, 0.625 inch thick, not just to allow some room for milling and gaining cc, but also to ensure good gasket life. The bolt holes for the cylinder head are all blind - shut off at the bottom so coolant will not get in and corrode the bottom of the bolt or stud threads. The bolt bosses in this block are deep and have good tie in with the deck and the cylinder walls.

Deck height refers to the distance from the centerline of the crank to the deck of the block - top of the cylinders. This all important dimension, together with the stroke, controls how long a rod and how tall a piston you can fit into the engine. The Rocket blocks are available with a standard 9.025 inch deck height and 350 main bearing bore size of 2.45 in either dry sump (22551657) or in wet sump style (22551788). Your other option is to order the tall 9.325 deck height, which gains 0.300 extra room for a longer rod and improved stroker action. Those blocks carry the larger 2.65 main bearing size. Here, our choice is a tall deck dry sump with the unmachined rear main (22551659) or the tall deck wet sump (2251790). The tall decks accept 6.000 inch rods as part of their natural birth right. You reduce the angle at which the rod operates, cut down on the side thrust at the piston and gain more top end power. The tall deck pulls rank in other ways. For instance the stock deck Rocket has the bores machined at 4.000 inch, while the tall deck in wet or dry sump comes with 4.125 bores.

EXPANSION ROOM

Normally, adding a stroker calls for milling the pan rails so the rods can

There are both front and rear 1/2 inch bosses for oil connections directly to the main oil gallery. This allows convenient connections for wet sump use.

There are two lifters galleries above the cam and a main gallery to the right of the cam. The oil is brought from the main gallery to the top of the main and then back up to the cam bearing.

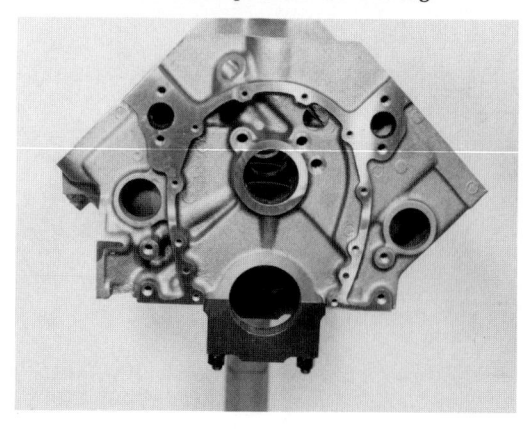

94 • 457 SMALL BLOCK

The steel center main bearing caps extend the full width of a block and have screwdriver slots which make it easier to pull them up. This rear main is for a dry sump engine. Heavy duty oil pan rails are 0.800 inch further apart.

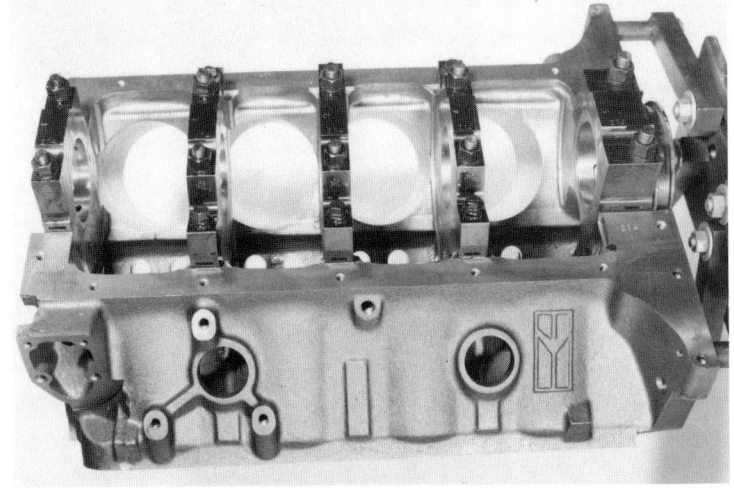

The rocket block is the most advanced small block currently available in cast iron. Ribbed outer walls add both strength and cooling space. Fuel pump pushrod is longer. Both dry and wet sump rear main caps are available.

clear. This works but leaves weakened pan rails and insufficient sealing surface at the pan. The Rocket block has massive pan rails and they are moved further apart than on a stock block by 0.800 inch. Now you can add a 4.00 inch stroker crank and the rods clear. With just a 4 inch crank and a tall deck, you go directly without stopping to assemble 427 cubic inches of Chevy power, which puts cost below that of the machined Bow Tie and gives you more cubic inches for the money plus extra strength. Among the little details involved in the wider block is that the fuel pump push rod is now 0.190 inch longer than stock.

BODY BUILDER

One look at the water jackets on the outside of the block, or for that matter at the water jacket walls in the lifter valley, reveal a ripped look. Nick Fowler explained to us that the cylinder bores were moved out and so was the water jacket. The outer water jacket walls just follow the cylinder wall contours to maintain even coolant flow. Also, the casting contours add strength.

Think of it this way: all power in the engine comes from the pressure on the pistons. Now all eight pistons take turns trying to tear the crank out of the main bearing webs. No wonder engines develop cracks at the main bearings or pound out the caps. With not enough support, the crank also lets go.

Here the front and rear bulkheads have been thickened substantially. More metal has been added to the top of the webs, just under the cylinders. Normally when the cylinder block is honed, an inch of overstroke is allowed for the honing head. (The hone goes an inch past the bottom of the bore.) Here the overstroke room of the hone is cut back to a half inch and the main webs are that much taller and stronger.

The steel billet center main bearing caps now extend across the block from side to side. Small screwdriver slots at each end help pry up the cap without hammering. Splayed outer bolts engage directly into the thickened pan rails. In a way it is like having a crossbolted block without the weight penalty. All of the Rocket blocks accept the two piece rear seal and the good guy old style cranks.

There are also dual starter mounts so you can install a starter on the left or right depending on clearance problems or the pan you select. Mounting pads and threaded holes are available on both sides, so you can change things later. This allows you to run a full length oil pan kick out and mount a stock sized the starter out of the way. The combination of wide pan rail spacing, the option of better kickouts and the starter mounting brings you to a new world of pans. Moroso makes custom pans oil pans for it and has Rocket blocks in stock, specially for pan fitting. The same goes for many other oil pan makers, Hamburger, Stef's, Milodon and more.

As part of the general expansion program, the cam has moved up 0.400 inch. This really uncorks the Rocket and gets it ready for takeoff. You have room for Oliver rods with a 3 3/4 crank, you can clear a Carillo rod with minimal 0.040 machining and you no longer need special cams with a small base circle. Cloyes makes a roller timing

The cam centerline is moved up a full 0.400 inch which gives you ample room for strokers with minimum machining. The oiling system is just like on the Olds Pro/Stock.

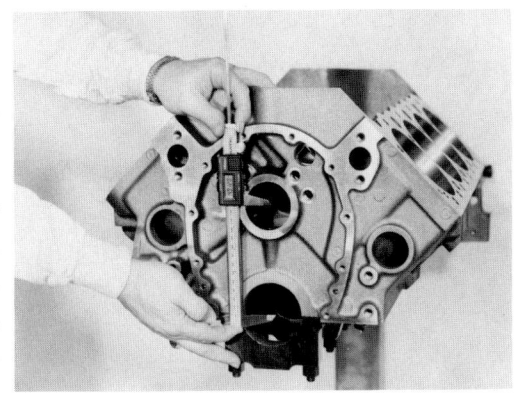

457 SMALL BLOCK • 95

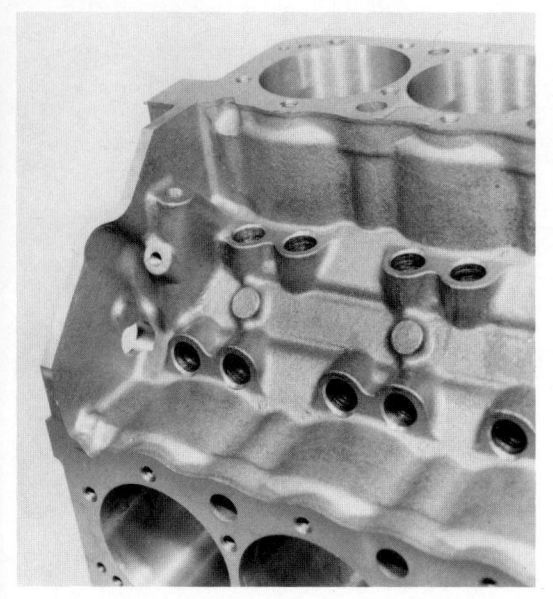

The front wall carries a large boss which leads directly to the main oil gallery. All cylinder head bosses are longer, for better tie-in. The bosses are also blind which eliminates contact with coolant; you won't need any sealers.

chain set for it and both Jessel and Bo Laws are developing timing belt systems.

Another step in the right direction was the switch from small block to big block cam journals. It achieves sturdier cam bearings, allows a physically larger cam with more lift or base circle and helps with giant valve springs. The increase in cam bearing bore size also improves the cam's torsional stiffness. As you can well expect, all cam manufacturers have cores in stock for the Rocket block.

The cam bearings have the big block size and this combined with moving up the cam allows a full sized base circle, even with high lifts. Cam bores in the block have no oil grooves and are also line honed for perfect alignment.

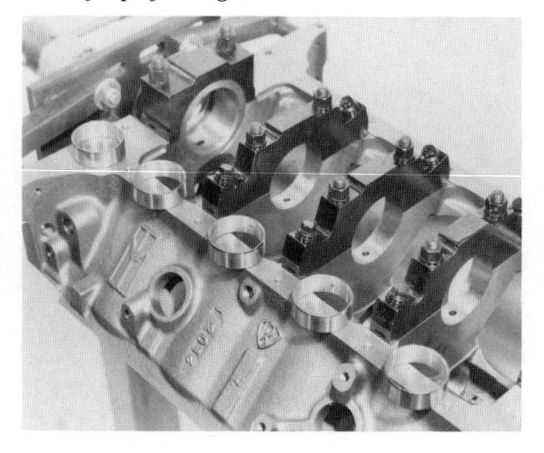

RACE OILING

None of the cam shaft bearing bores uses the stock annular grooves to feed oil around the cam bearing. This achieves a number of benefits, first of which is that the cam bearings will not spin. They have better support and receive a stronger grip from the block. Unlike stock cam bores, these are all of the same size and are line honed in production. Now you can throw away the bearing scrapers and stop wondering about oil pressure loss.

The new oiling system is very similar to the Olds Pro/Stock. If you look at the front of the block, there are two oil lifter galleries just above the cam bearing bore at 11 and 1 o'clock and the main gallery at 3 o'clock. Oil from the main gallery is directed through an angled channel to each main bearing. Oil then goes from the top of the main to the bottom of the cam bearing through a separate vertical gallery. Mains get first call on oil directly from the main oil gallery.

You will not find an oil filter pad on this engine. All connections are external - there is one on the side of the block, another at the front and a a third at the rear - so you can oil a sprint car, late model or drag machine with an external filter and the connections you want. It also gets the filter out of the way of the starter and the headers and eliminates a bunch of right angle turns from the system.

You have a choice of wet sump or dry sump oiling. On the wet sump, the pump is moved down 0.300 to accommodate larger strokes and uses the big block oil pump drive. The wet sump block has a rear main cap machined to match, while the dry sump cap is solid.

Cam oiling is an interesting point. On a small block Chevy, the smart guys introduce cam oiling at 3 or 4 o'clock, looking at the front of the bearing. The cam drags this oil toward the bottom point which leaves an oil wedge in the high load bottom area. If you introduce the oil from the bottom as on the Rocket block, it may become desirable to provide the ID of the bearing with a tail groove extending from the 6 o'clock entry toward 9 o'clock.

One oddity is that the distributor is now tipped at 5.0 degrees instead of 4.5 degrees. With digital CNC controls, it is easier to set up the machine that way. Spot face the intake manifold distributor seat to match to gain a good seal.

Rockets comes with a space age price tag, not much change from 2300 dollars but when you compare it to other blocks of equivalent strength, things get interesting. Yes, you can get cast iron blocks from 700 on up. By the time you install steel mains on a two bolt block, you do have a better piece but the price tag is up there. When you get a respectable Bow Tie with splayed cap, the initial cost is 1400 dollars and you still have to bore it and machine it.

The Rocket comes fully machined with 4.000 bores on stock deck heights and has the bores opened to 4.125 on the tall decks wet or dry sump. The tall decks accept 6.000 inch rods as part of their natural birth right.

Many cubic inch combinations are available to you, ranging from 350 to 400. From there, a 4.125 bore by 4.000 inch stroke will give you 427 cubic inches. Using a 4.200 bore and the same 4.00 inch crank delivers 443 cubic inches. If you switch to a 4.125 crank, you reach 457 cubic inches without exceeding the capability of the block. In that light, the Rocket has orbited around the Bow Tie and is excellent horsepower value.

There are bosses for optional stand pipes or plugs. Note the extensive block wall convolutions which help stiffen the structure.

96 • 457 SMALL BLOCK

FACT FINDER

We stopped off for a visit with engine builder Ross Lombardi in Niles, OH and came up with a handy reference list of Chevy facts.

OIL FILTER

Oil up the gasket, prefill the filter, so you don't start the engine dry and tighten hand tight so the filter gasket just contacts. Tighten 3/4 of a turn beyond the initial tightening point. Overtightening can fracture an oil filter and kill the engine. Tightening with pliers or anything else that kinks a filter will cause it to fail. Always check that there are no leaks before driving a car with a new filter.

SPARK PLUGS

Check that the thread reach corresponds to your engine. Consult your Champion catalog for the correct heat range. You generally do not have room for a torque wrench. For gasketed plugs, Champion suggest tightening the plug finger tight and then going one quarter turn further to seat it. Overtightening a plug changes its heat range and can fracture the insulator. With a taper seat plug tighten 1/16 of a turn past the finger tight position. Torque for a gasketed 14 mm. plug in cast iron is 26–30 lb.-ft., in an aluminum head with a gasket seat 18–22 lb.-ft., with a tapered seat tighten 7–5 lb.-ft. Always use anti-seize on plug threads in aluminum. Plug threads are easy to strip out. Plug repair inserts are available from Helicoil or Time Sert.

VALVE LASH

Mechanical lifters: check the valve lash specified by your manufacturer on the timing card. You can gain a little extra duration by closing up on the valve lash, but you can also burn up the valves when they are set too tight. When you run into a miss, check for a tight valve, in addition to changing the plugs. If you have a click with a mechanical lifter, try tightening the clearance a little or test for a sound change with a feeler gauge.

Hydraulic lifters: on a new engine with no manifold, take out the lash till the pushrod just turns but has not pushed down on the plunger, so that it is still up against the spring clip. Then tighten one turn. On a race engine with hydraulic lifters lash out the lifters by tightening only a quarter turn. On a street engine with miles, have the engine running and warmed up, and tighten till the noise is out. To prevent an oil bath use anti-spray clips or cut down valve covers where the top is off for access. Tight valves will cause a miss and will burn.

LIFTER BORES

A quick clean up with a ball hone or flapper paper. Use a 0.001 to 0.002 clearance in cast iron. With not enough clearance they hang up.

With bronze bushings and roller lifters, run 0.001.

MAIN BEARINGS

Use a 0.0025 to 0.0026 clearance—nothing over 0.003. You can see a 20 psi oil pressure change between 0.0025 and 0.003.

CRANK END CLEARANCE

Shoot for 0.005 to 0.006. More end clearance hurts the timing chain. The crank and cam sprockets must be aligned.

ROD BEARINGS

Rod bearing clearance should be 0.002 to 0.003 with 0.0025 as a happy medium. More clearance throws more oil and now you need tighter rings—that is not a gain.

ROD SIDE CLEARANCE

Go for 0.015 to 0.025

PIN BORE

0.0008 with gas and 0.0012 on alcohol

RING GAP

Top Ring 0.018—0.020
Second Ring 0.016—0.018

PISTON SIDE CLEARANCE

From your piston spec sheet find out how they are measured. Usually across the thrust face at level of the wrist pin.

Silicon forging: 0.005 to 0.006—silicon has good expansion control, as much as 0.007–0.008 on some.

Hypereutectic 0.004 to 0.005.

PISTON TO INTAKE VALVE

0.075 to 0.100

PISTON TO EXHAUST VALVE

0.100 to 0.150

CAM BEARINGS

0.002 to 0.004—they do have an effect on oil pressure.

CAM END CLEARANCE (Thrust Button)

0.003 to 0.004—you need some oil or the block and the button fail.

OIL PAN SCREEN TO OIL PAN:

0.250

Newest from Chevrolet is the Corvette LT5 with two cams in each head, four valves per cylinder, and wet cylinder liners in the aluminum block. Centrally located plug insures fast flame propagation.

FIRING ORDER

1-8-4-3-6-5-7-2

TORQUE

Here are some ballpark figures. They are given for oiled threads; if you use moly lube, the torque figures have to be reduced.

Heads 70 ft.-lb.
Mains 80 ft.-lb.
Flywheel 65—70
Clutch 30 -35

Rods It depends entirely on the rod bolts you are using. A 22 dollar SPS bolt calls for a 90 ft.-lb., while a WMD 7 calls for 70 ft.-lb.

Stock rods 45 ft.-lb.
Exhaust manifold bolt 19 ft.-lb.

The LT5 is Chevrolet's most advanced engine and the Raceshop is already busy planning race-winning extras for it.

FACT FINDER • 97

ABS PRODUCTS
9534 Atlantic Ave. P.O.B.1984
South Gate, CA 90280
213 564-9076

ARP
8565 Canoga Avenue
Canoga Park, CA 91304
1 800 826 3045
818 341-4488

B-H-J INC.
37530 Enterprise Court,#3
Newark, CA 94560
415 797-6780

BLP PRODUCTS, INC.
2545 Industrial Boulevard
Orlando. FL 32804
407 294-4330

BRODIX, INC.
P.O.Box 1347
Mena, AR 91953
501 394-1075

CALLIES PERFORMANCE PRODUCTS
P.O. Box 670 A
Fostoria, OH 44830
419 435-2711

CHILDS AND ALBERT, INC.
24849 Anza Drive
Valencia, CA 91355
805 295-1900

DIAMOND RACING PRODUCTS, INC.
23003 Diamond Drive
Mt.Clemens, MI 48043
313 792-6620

EDELBROCK CORP.
2700 California St.
Torrance, CA 90503
213 781-2222

EVANS SPEED EQUIPMENT
2550 Seaman
South El Monte, CA
213 444-2838

FEL-PRO INC.
7450 North McCormack Blvd.
Skokie, IL 60076-8103
312 674-7700

FLUIDAMPR - VIBRATECH, INC.
537 East Delavan Avenue
Buffalo, NY 14211
716 895-8000

GAERTE ENGINES
Rochester, IN
219 223-3016

HOWELL ENGINE DEVELOPMENTS
24356 Sorrentino Court
Mt. Clemens, MI 48043
313 791 6400

JEGS
751 East 11th Avenue
Columbus, OH 43211
614 294-5454

JESEL
P.O. Box 1407
Wall, NJ 07719
201 681-5344

KATECH
22969 Rasch.
Mount Cclemens, MI 48043
313 791-4120

LAB ENGINE MACHINE
128 Albany Ave.
Lindenhurst, NY 11757
516 226-2241

MC PHERSON CHEVROLET
21 Auto Center Drive
Irvine, CA 92718
714 768-7222

MAYFAIR AUTOMOTIVE MACHINE
20419 VanBorn Rd.
Taylor, MI 48180
313 561-1497

MR. GASKET CO.
8700 Brookpark Rd.
Cleveland, OH 44129
216 398-8300

BILL MITCHELL PRODUCTS, INC.
35 Trade Zone Drive
Ronkonkoma, NY 11779
516 737-0372

MITUTOYO
18 Essex Rd.
Paramus, NJ 07652
201 368-0525

MOLDEX CRANK SHAFTS
25229 Warren Ave.
Dearborn Heights, MI 48127
313 561-7676

MOROSO
80 Carter Drive
Guilford, CT 06437
203 453-6571

OLIVER RODS
1025 Clancey Ave., N.E.
Grand Rapids, MI 49503
1 800 253-8108
616 451-8333

PMS (PRECISION MEASUREMENT)
P.O. Box 28097
San Antonio, TX 78228
512 681-2405

POWERFORMANCE INTERNATIONAL
2225 West Mountain View #17
Phoenix, AZ 85021
602 242-9421
1 800 874-2573

R.H.S.
3410 Democrat, Rd.
Memphis, TN 38118
901 794-2832

RODECK ALUMINUM BLOCKS
18093 S. Figueroa
Gardena, CA 90248
213 538-5791

SCOGGIN-DICKEY
5901 Spur 327
P.O.B. 64910
Lubbock, TX 79464
1 800 456-0211

SEAPORT AUTOMOTIVE
2607 W. Central Ave.
Toledo, OH 43606
419 474-0519

SHAVER SPECIALTIES
20608 Earl St.
Torrance, CA 90503
213 370-6941

SPEED-PRO
100 Terrace Plaza
Muskegeon, MI 49443
616 724-5011

STAHL HEADERS
1515 Mt. Rose Ave.
York, PA 17403
717 846-1632

HOWARD STEWART ENGINE
COMPONENTS
P.O. Box 5523
High Point, NC 27262
919 889-8789

SUMMERS BROTHERS INC.
530 S. Mountain Avenue
Ontario, CA 91762
714 986-2041

SUNNEN PRODUCTS
7910 Manchester
St. Louis, MO 63143
314 781-2100

VANDERVELL AMERICA, INC.
2488 Tuckerstone Pkwy.
Tucker, GA 30084
1 800 241-3498
404 491-3935

WEAVER BROS. LTD.
1980 Boeing Way
Carson Ccity, NV 89706
702 883-7677

WEIAND MANIFOLDS
P.O.Box 65977
Los Angeles, CA 90065
213 255-4138